Blockchain for IoT Systems

Blockchain and distributed-ledger technologies enable new modes of communication, synchronization, and transfer of value with a broad impact on Internet of Things (IoT), Data Science, society, industry, commerce, and government. This book studies the potential impact of blockchain and distributed-ledger technologies on IoT. It highlights the application of possible solutions in the domain of blockchain and IoT system security, including cryptology, distributed systems, law, formal methods, code verification and validation, software, and systems metrics. As the field is growing fast, the book adapts to the changing research landscape, integrates and cross-links studies and citations in related subfields, and provides an overview of these fields and how they complement each other.

- Highlights how the security aspect of the integration of blockchain with IoT will help to design secure data solutions for various domains
- Offers fundamental knowledge of the blockchain concept and its usage in real-life applications
- Presents current and future trends on the IoT and blockchain with an efficient, scalable, and sustainable approach
- Reviews future developments in blockchain and IoT in future job opportunities
- Discusses how blockchain for IoT systems can help a varied range of end-users to access computational and storage resources

This book is intended for postgraduate students and researchers in the departments of computer science, working in the areas of IoT, blockchain, deep learning, machine learning, image processing, and big data.

Blockchain for Smart and Green Society: Promise, Practice and Application

Series Editors: Vishal Bhatnagar and Vikram Bali

List of titles:

Blockchain for Smart Systems
Computing Technologies and Applications
Edited by Latesh Malik, Sandhya Arora, Urmila Shrawankar and Vivek Deshpande

Blockchain for IoT Systems
Concept, Framework and Applications
Edited by V. Sridhar, Sita Rani, Piyush Kumar Pareek, Pankaj Bhambri and Ahmed A. Elngar

For more information about this series please visit Chapman & Hall/CRC Blockchain for Smart and Green Society: Promise, Practice and Applications—Book Series—Routledge & CRC Press

Blockchain for IoT Systems

Concept, Framework and Applications

Edited by
V. Sridhar, Sita Rani, Piyush Kumar Pareek,
Pankaj Bhambri and Ahmed A. Elngar

CRC Press
Taylor & Francis Group
Boca Raton London New York

CRC Press is an imprint of the
Taylor & Francis Group, an **informa** business
A CHAPMAN & HALL BOOK

Designed cover image: ShutterStock

First edition published 2025
by CRC Press
2385 NW Executive Center Drive, Suite 320, Boca Raton FL 33431

and by CRC Press
4 Park Square, Milton Park, Abingdon, Oxon, OX14 4RN

CRC Press is an imprint of Taylor & Francis Group, LLC

ISBN: 978-1-032-60708-5 (hbk)
ISBN: 978-1-032-60740-5 (pbk)
ISBN: 978-1-003-46036-7 (ebk)

DOI: 10.1201/9781003460367

Typeset in Times LT Std
by Apex CoVantage, LLC

Contents

Preface

This book is designed to illustrate the role of blockchain technology in improving IoT applications and systems by offering a decentralized, safe ledger for open, unchangeable data exchange among devices. In order to explore a dependable and effective ecosystem for blockchain and IoT aspirants, a variety of applications ranging from smart homes to industrial automation are included.

While working on this book, special emphasis was given to present blockchain solutions for real-life IoT deployments. Major challenges in deploying these solutions are also presented along with possible future research directions. So, this book caters to the needs of IoT engineers, blockchain professionals, academia, research scholars, and industry.

The areas of strength of this book are as follows:

- It provides a comprehensive coverage of the subject in a very lucid manner.
- Chapters from researchers working in this domain in various countries are included to provide a global exposure.
- It provides a plenty of real-life examples and case studies for in-depth knowledge of the readers.

We are certain that readers will enjoy reading this book and it will go beyond their expectations.

Editorial Team

Editors

Dr. V. Sridhar is currently serving at Nitte Meenakshi Institute of Technology, Bengaluru, as Professor and Dean of Academics. His educational qualifications include B.E. (PESCE, Mandya, University of Mysore), M.E. (Jadavpur University, Calcutta), Ph.D. (Indian Institute of Technology, New Delhi), and Postdoctoral Research (Department of Electrical & Electronics, University Tenaga Nasional [UNITEN], Selangor, Malaysia, from November 27, 2000, to November 26, 2002). He has over 40+ years of vast experience and has to his credit 44+ documents indexed in Scopus with H-Index of 7+. His major fields of interest are Routing Protocols, Wireless Sensor Networks, and Internet of Things. He has awarded Ph.D. to more than eight research scholars. He has attended several conferences and Institutional Interaction programs at the United States, Canada, Malaysia, Singapore, Thailand, Indonesia, France, and Dubai. He has successfully carried out more than six funded projects.

Dr. Sita Rani works in the Department of Computer Science and Engineering at Guru Nanak Dev Engineering College, Ludhiana. She earned her Ph.D. in Computer Science and Engineering from I.K. Gujral Punjab Technical University, Kapurthala, Punjab, in 2018. She has also completed Post Graduate Certificate Program in Data Science and Machine Learning from Indian Institute of Technology, Roorkee, in 2023. She has completed her Postdoc from Big Data Mining and Machine Learning Lab, South Ural State University, Russia, in August 2023. She has more than 20 years of teaching experience. She is an active member of ISTE, IEEE, and IAEngg. She is the recipient of ISTE Section Best Teacher Award-2020 and International Young Scientist Award-2021. She has contributed to the various research activities while publishing articles in the renowned SCI and Scopus journals and conference proceedings. She has published seven international patents and authored, edited, and coedited nine books. Dr. Rani has delivered many expert talks in AICTE-sponsored Faculty Development Programs and key note talks in many national and international conferences. She has also organized many international conferences during her 20 years of teaching experience. She is the member of Editorial Board and reviewer of many international journals of repute. She is also the vice president of SME and MSME (UT Council), Women Indian Chamber of Commerce and Industry (WICCI), since last three years. Her research interest includes Parallel and Distributed Computing, Data Science, Machine Learning, Internet of Things (IoT), and Smart Healthcare.

Dr. Pankaj Bhambri works in the Department of Information Technology at Ludhiana's Guru Nanak Dev Engineering College, Ludhiana. He serves as the Institute's Coordinator, Skill Enhancement Cell, and has almost two decades of teaching experience. He earned his M.Tech. (CSE) and a B.E. (IT) with honors from the I.K.G. Punjab Technical University, Jalandhar, and Dr. B.R. Ambedkar University, Agra, India, respectively. Dr. Bhambri earned his doctorate in computer science and engineering from the I.K.G. Punjab Technical University, Jalandhar, India. His research has appeared in a variety of prestigious international/national journals and conference proceedings and he has contributed to numerous books and has also filed several patents. Dr. Bhambri has been awarded the ISTE Best Teacher Award in 2023 and 2022, the I2OR National Award in 2020, the Green ThinkerZ Top 100 International Distinguished Educators in 2020, the I2OR Outstanding Educator Award in 2019, the LCHC Best Teacher Award in 2007, the CIPS Rashtriya Rattan Award in 2008, and the SAA Distinguished Alumni Award in 2012 along with countless other accolades from various government and non-profit organizations. Machine Learning, Bioinformatics, Wireless Sensor Networks, and Network Security are his areas of interest.

Dr. Piyush Kumar Pareek is currently serving as Professor and Head of the Department of Artificial Intelligence and Machine Learning & Intellectual Property Rights Cell of Nitte Meenakshi Institute of Technology, Bengaluru. He has interest in continuous learning and teaching. He completed his B.E., M.Tech., Ph.D., and Post Doc in the field of Computer Science Engineering, Registered Patent Agent, Government of India. He has 12+ years of experience in teaching. He has to his credit 100+ research articles published in Scopus Indexed journals/conferences; 10+ textbooks; 50+ Industrial Design, registered at Indian Patent Office; 25+ International Patents granted; and 50+ published Indian Utility Patents. He has guided Ph.D. scholars in Visvesvaraya Technological University, Belagavi, University of Mysore, Manipal University Bengaluru, and Nitte University. He has awarded Ph.D. to six students in Visvesvaraya Technological University, Belagavi. He is Senior Member of IEEE and MIE. He is Guest Editor of *MDPI Electronics* and Taylor & Francis Book Series and Reviewer in Springer, Elsevier, Wiley, and Hindawi journals. He is also IPR and Publications Advisor for the Education Sector.

Dr. Ahmed A. Elngar is the Founder and Head of Scientific Innovation Research Group (SIRG) and Assistant Professor of Computer Science at the Faculty of Computers and Information, Beni-Suef University. Dr. Elngar is a Director of the Technological and Informatics Studies Center (TISC), Faculty of Computers and Information, Beni-Suef University, Egypt. He is a Managing Editor of the Journal of Cybersecurity and Information Management (JCIM). Dr. Elngar has more than 25 scientific research papers published in prestigious international journals and over five books covering such diverse topics as data mining, intelligent systems, social networks, and smart environment. He is a member of the Egyptian Mathematical Society (EMS) and International Rough Set Society (IRSS). His other research areas include the Internet of Things (IoT), Network Security, Machine Learning, and Artificial Intelligence. He is an Editor and Reviewer of many international journals around the world. Dr. Elngar won several awards including the *Young Researcher Award in Computer Science Engineering* from Global Outreach Education Summit and Awards 2019. He is also a recipient of *the Best Young Researcher Award (Male) (Below 40 years)* from Global Education and Corporate Leadership Awards (GECL-2018). Dr. Elngar has many activities in the community and the environment service including organizing 12 workshops hosted by a large number of universities in almost all governorates of Egypt.

Contributors

Mamatha A.
Ramaiah Institute of Technology
India

Mahfooz Alam
Aligarh Muslim University
Aligarh, India

Shilpa Ankalaki
Manipal Academy of Higher Education
Manipal, India

Anwesha Banik
GH Raisoni College of Engineering and Management
Pune, India

Radia Belkeziz
Private University of Marrakesh
Marrakesh, Morocco
and
Laboratory of Computer and Systems Engineering
Cadi Ayyad University
Marrakesh, Morocco

Pankaj Bhambri
Guru Nanak Dev Engineering College
Ludhiana, India

Sachin Chaudhary
IIMT University
Meerut, India

Reda Chefira
Private University of Marrakesh
Marrakesh, Morocco
and
Laboratory of Computer and Systems Engineering
Cadi Ayyad University
Marrakesh, Morocco

Sudarsanan D.
Cambridge Institute of Technology
Bengaluru, India

Santosh L. Deshpande
Visvesvaraya Technological University
Belagavi, India

Madhura G.K.
Nitte Meenakshi Institute of Technology
Bengaluru, India

Dhananjaya G.M.
Visvesvaraya Technological University
Belagavi, India

Geetabai S. Hukkeri
Manipal Academy of Higher Education
Manipal, India

Anita Kanavalli
Ramaiah Institute of Technology
Bengaluru, India

Aman Kataria
Amity University
Noida, India

Manjunath R. Kounte
REVA University
Bengaluru, India

Bhupendra Kumar
IIMT University
Meerut, India

Shashank Kumar
Sharda University
Noida, India

V. Sunil Kumar
Nitte Meenakshi Institute of Technology
Bengaluru, India

Aejaz Nazir Lone
Aligarh Muslim University
Aligarh, India

Santosh M.
Cambridge Institute of Technology
Bengaluru, India

Vaneeta M.
Ramaiah Institute of Technology
Bengaluru, India

Vasantha M.
Cambridge Institute of Technology
Bengaluru, India

Waseem Ahmad Mir
GH Raisoni College of Engineering and Management
Pune, India

Suhel Mustajab
Aligarh Muslim University
Aligarh, India

Iqra Nissar
Jamia Millia Islamia
New Delhi, India

Bharati B. Pannyagol
Visvesvaraya Technological University
Belagavi, India

Chandra Shekar Pant
Government PG College
Ranikhet, India

Rajneesh Panwar
IIMT University
Meerut, India

Piyush Kumar Pareek
Nitte Meenakshi Institute of Technology
Bengaluru, India

Vikram Puri
Duy Tan University
Da Nang, Vietnam

Chandramma R.
Global Academy of Technology
Bengaluru, India

Said Rakrak
Laboratory of Computer and Systems Engineering
Cadi Ayad University
Marrakesh, Morocco

Sita Rani
Guru Nanak Dev Engineering College
Ludhiana, India

Somashekhara Reddy D.
Jain University
Bangalore, India

Renukadevi S.
Jain University
Bengaluru, India

Mohamed Yassine Samiri
Laboratory of Computer and Systems Engineering
Cadi Ayad University
Marrakesh, Morocco

Mohan S. G.
Nitte Meenakshi Institute of Technology
Bengaluru, India

Udhayaranjani Sellappagounder Mohan
Dalhousie University
Halifax, Canada

Mohan S.G.
Nitte Meenakshi Institute of Technology
Bengaluru, India

Tawseef Ayoub Shaikh
National Institute of Technology
Srinagar, India

Kewal Krishan Sharma
IIMT University
Meerut, India

Vikas Sharma
IIMT University
Meerut, India

K. Aditya Shastry
Nitte Meenakshi Institute of Technology
Bengaluru, India

V. Sridhar
Nitte Meenakshi Institute of Technology
Bengaluru, India

Rishik Srivastava
Sharda University
Noida, India

Md. Tauseef
REVA University
Bengaluru, India

Sangeetha V.
Ramaiah Institute of Technology
Bengaluru, India

Tarun Kumar Vashishth
IIMT University
Meerut, India

Vani Vasudevan
Nitte Meenakshi Institute of Technology
Bengaluru, India

Priti Verma
Sharda University
Noida, India

1 Blockchain Technology
Applications and Challenges

Priti Verma, Rishik Srivastava, and Shashank Kumar

1.1 INTRODUCTION

Blockchain technology has evolved beyond its original use and is now a transformational force with broad repercussions across businesses and sectors. It was initially developed as the foundational infrastructure for cryptocurrencies like Bitcoin. Blockchain is a decentralized, transparent, and irreversible ledger system at its heart that has the ability to completely change how data and value are handled, moved, and safeguarded. This chapter digs into the complex world of blockchain, examining its numerous uses and the enormous obstacles it must overcome to achieve general adoption (Bali et al., 2023). The blockchain technology, which was initially associated with cryptocurrency only, has emerged as a powerful and revolutionary technology for lucrative business in multiple areas. Practitioners from diverse sectors have joined an open conversation about this technology's potential disruptive possibilities. According to recent projections, blockchain will hold about 10% of the world's gross domestic product (GDP) by 2025 (Rani et al., 2023a). Nevertheless, because of the intricacy of technology and complex issues in using it across multiple application domains, the scientists, researchers and users find it difficult to predict a single vision of the future of blockchain technology. Still, there are basic questions that need to be addressed. For example, despite significant and undisputable technological innovation associated with this technology, there exist doubts and conjectures about the possible outcomes of its use. The answers to questions, like which applications will be implemented or what will be the eventual societal impacts of these changes, still aspire for clarity.

Blockchain technology has erupted as a transformative force with enormous potential to alter the future of many industries. Blockchain, which was first associated with cryptocurrencies such as Bitcoin, has grown into a diverse and powerful platform that offers much more than simply digital money. Blockchain is set to disrupt established systems and offer up new avenues for innovation and efficiency because of its key principles of decentralization, transparency, immutability, and security.

Blockchain, as a decentralized platform and distributed ledger database, is supposed to promote the centralized system to shape an all-inclusive ecosystem. It is expected that by the end of the third decade of the 21st century, it will generate easy, effective, efficient, and faster methodologies to exchange value and data information among users, organizations, and nations across the world. New opportunities

DOI: 10.1201/9781003460367-1

will explode, and this technology will boost and unravel the innovative potential latent within companies and societies. One essential prerequisite will be to frame policies for the concerned sectors to enable them to create and develop a competitive edge at the global platform (Kumar et al., 2022). From this perspective, the involvement and competence of central governments to encourage blockchain technology and to boost innovation with flexible policies will be a crucial driving force to implement this technology and boost productivity and growth throughout the world. This chapter seeks to give a complete knowledge of blockchain technology's uses and the complicated web of difficulties that determine its development through an examination of real-world examples, industry case studies, and expert commentary. As we go through the many industries and use cases, we will discover the transformational potential of blockchain technology while also recognizing the obstacles that must be overcome in order to fully realize its promise. As a result, we begin on a fascinating investigation of one of the most revolutionary technical advances of our time.

1.2 APPLICATIONS AND CHALLENGES IN VARIOUS INDUSTRIES

1.2.1 Financial Services Industry

The financial sector was the first one to use blockchain technology for some financial services, notably for uses like cryptocurrency and smart contracts. However, there are still a number of issues. Scalability, in particular, continues to be a critical concern (Rani et al., 2023a). The majority of blockchain networks, including Bitcoin and Ethereum, include throughput restrictions on transactions. Due to congestion and high transaction costs during peak hours, widespread adoption has been hampered. Additionally, regulatory worries have hampered the development of blockchain in finance (Rani et al., 2023b). Different nations have taken varying positions on regulating cryptocurrencies, making the business environment difficult. To promote confidence and promote innovation in this industry, clear, uniform, and universally understood laws are essential.

Additionally, there are still several issues with interoperability between various blockchain networks and established financial institutions. For blockchain to realize its promise in sectors like cross-border payments, securities trading, and supply chain financing, these problems must be resolved.

1.2.2 Supply Chain Management

Blockchain technology promises to make supply chains transparent and traceable, which will cut down on fraud, counterfeiting, and inefficiencies. However, there are a number of obstacles to its acceptance in this sector. Integration of current systems with blockchain networks is one of the main issues (Bhambri & Rani, 2024). The change is difficult and expensive since many supply chain actors utilize outdated systems that are incompatible with blockchain technology. Additionally, it is difficult to guarantee data validity and integrity on the blockchain. The accuracy of the data submitted into the system cannot be guaranteed by blockchain, despite the fact that it can offer immutable records. As a result, strong standards and data validation processes are required to ensure the accuracy of the data documented on the blockchain.

In supply chain applications, scalability and cost-effectiveness are additional problems. Large datasets across the whole supply chain can be sluggish and resource-intensive to store on a blockchain, possibly compromising the efficiency improvements that the technology promises.

1.2.3 Healthcare Sector

Electronic health records (EHRs) and medication tracing show enormous potential for safeguarding and simplifying healthcare data management, thanks to blockchain technology. However, there are several difficulties in this industry. Due to the sensitivity of healthcare data, privacy considerations are of utmost importance. It can be difficult to maintain compliance with laws like Health Insurance Portability and Acountability Act (HIPAA) and strike a balance between transparency and privacy. Another key issue is interoperability. Healthcare systems throughout the world employ a diverse set of proprietary software and standards, making it challenging to easily incorporate blockchain technologies. This dispersion makes it difficult to build a cohesive, interoperable healthcare blockchain ecosystem.

Furthermore, due to the requirement for training and infrastructure changes, enrolling healthcare providers onto a blockchain network can be time-consuming and costly. Scalability is also an issue, especially when dealing with the massive amounts of data created in the healthcare industry.

1.2.4 Agriculture and Food Sector

Blockchain technology has been investigated by the food and agricultural industry to improve food safety and traceability. However, it has problems with data reliability and trust. It can be challenging to ensure that information on the origin and quality of food goods uploaded into the blockchain is correct, especially in supply chains that are complicated and involve several parties.

The expense and difficulties of installing blockchain at different points along the supply chain, from farmers to consumers, provide another difficulty. Particularly small-scale farmers might not have the infrastructure or resources needed to properly utilize blockchain technology. Furthermore, interoperability becomes a problem when several blockchain networks are utilized by various food producers or in various geographical locations. To achieve widespread acceptance, a single and uniform framework for food traceability must be developed Puri et al. (2022).

Blockchain is a kind of peer-to-peer distributed ledger which is shared and synchronized among computers. Each and every computer preserves its own set of records that constitute blocks in the blockchain. A nonce, the cryptographic hash of the preceding block, and an accumulation of transactions constitute each block. Nonces are random numbers that are used to validate hashes. Each block contains the hash of the preceding block, therefore the links between them are made by the hash pointer. To capture the precise creation time of each record, the blocks are timestamped. The expression "blockchain" relates to the chain of blocks established by these interconnected blocks.

A blockchain network doesn't need a central authority given that each node includes a record of historical data and information of every single transaction, and provides it access to information at any point in time. Through this, the network of users can validate the transactions without the assistance of a trustworthy third-party mediator. For this they use a decentralized consensus mechanism before adding them in order to create a block. A transaction can no longer be changed after it has been integrated into a block because doing so requires amending all blocks that came before it. Even if an adjustment is effective in one node, the process must be performed in every node that is a participant, which uses a lot of processing power. As a result, blockchain has a tamper-proof feature and is secure by nature. It overcomes the centuries-old problem of resource duplication induced by employing it a number of times across multiple procedures. Asymmetric encryption, P2P, distributed ledger, consensus mechanism, and smart contracts are a few of the key technologies used in blockchain (Rani et al., 2023b).

A smart contract which is executed on a blockchain cannot be changed to alter since blockchain records are immutable and tamper-proof. Transferring the digital assets is rendered feasible through the automatic execution of predefined contract conditions, which guarantees that they are carried out exactly as agreed. Therefore, a blockchain's smart contract has the potential to provide protection from human discretional flaws and contract violations.

1.2.5 Education Sector

For tasks like confirming credentials and guaranteeing the integrity of educational data, blockchain technology has begun to gain traction in the education industry. However, issues with data ownership and control do exist. It can be difficult to determine if students, institutions, or outside sources own and manage educational records on a blockchain. The standardization of credentials on the blockchain to guarantee that they are widely acknowledged and accepted presents another difficulty. It is a difficult undertaking to reach agreement on such criteria among educational institutions and companies. When personal and academic data are housed on a blockchain, privacy and data security issues also surface. In this industry, it is crucial to strike a balance between transparency and data protection.

1.2.6 Legal Sector

By allowing smart contracts, automating legal procedures, and improving the security of legal documents, blockchain technology can bring a transformation in the legal industry. However, implementing blockchain technology in the legal sector presents special difficulties. One significant problem is the requirement to modify current legal frameworks to take into account smart contracts and blockchain-based proof. Decentralized and automated systems may make it difficult to apply conventional legal ideas. Additionally, it takes specialized knowledge to ensure the legal enforceability of smart contracts and resolve disputes that result from blockchain transactions. These issues may also call for modifications to arbitration and dispute resolution procedures. Blockchain can also increase the security of legal papers, but it also raises concerns about the long-term retention of legal information and the right to privacy in achieving balance (Rani et al., 2023c)

1.2.7 Energy Sector

Blockchain technology has been researched for uses in the energy sector, including peer-to-peer energy trade, grid management, and tracking carbon credits. However, the industry has unique obstacles. When dealing with the massive amounts of data created by energy networks, scalability is a major challenge. It is critical to ensure that blockchain can handle real-time transactions and data management at the scale demanded by the energy sector. Furthermore, energy rules and policies differ greatly among locations, complicating the application of blockchain solutions. Interoperability across multiple energy markets and grids is critical for realizing blockchain's full promise in this business. Furthermore, energy blockchain applications frequently necessitate the integration of IoT devices for data collecting and automation. It is difficult to ensure the security and dependability of these gadgets.

1.2.8 Manufacturing Sector

Today, the companies, factories, and organizations across the world are becoming increasingly interconnected; hence, the prevalence of blockchain becomes more relevant. The future manufacturing world spans across a network of products, machines, devices, suppliers, logistics companies, and what not. However, along with this environment of increasing network and virtual teams, today, more than ever before, security of data within and without the organizations is a big challenge for manufacturers.

To make the maximum use of this technology, the manufacturers can begin with self-analysis by identifying the issues and needs. Subsequently, it should be explored how the technology can address its needs and relieve the factory's pain points like transparent supply chain monitoring, asset tracking, and complex product designing.

The implementation of blockchain technology in the manufacturing sector can revolutionize the manufacturing sector. It can assist in multiple ways: streamlining operations, maintaining an effective supply chain, gaining greater visibility, and tracking assets and inventories with unprecedented precision. Blockchain has immense potential to transform the methods and process of designing, engineering, making, and scaling the products. By fostering trust among competitors within a common ecosystem, the technology is rewriting how competitors interact. Transparency and trust across all levels of the industrial value chain, from obtaining raw materials to supplying the finished product, can be scaled by blockchain.

Data-sharing and synchronization can be executed by distributed ledger technology (DLT), a decentralized method of data storage that relies on a network of many nodes and can boost the system performance while reducing the system maintenance costs simultaneously.

Smart contract program can self-execute, self-enforce, self-verify, and self-restraint. It operates on a blockchain platform and is deployed there. An agreement established between untrusted parties can have its predefined terms executed and enforced by using the code in a smart contract. To enforce an agreement's terms, smart contracts have the potential to work without reliable third-party intermediaries like banks and attorneys, which would result in lower transaction costs. Time is saved by automating processes without the use of a middleman (Rani et al., 2023d).

1.3 SMART MANUFACTURING THROUGH BLOCKCHAIN

Security of Data: Blockchain's chain storage structure with a timestamp, which is based on its capacity to protect information, allows smart manufacturing systems to operate. A timestamp along with a reference to the preceding block are included in each blockchain block. Data that has been recorded in a block cannot be changed once it has been done so. Data security is ensured by the encryption algorithm. Additionally, blockchain's traceability makes it easier to optimize production systems. In a real-world scenario, Industrial Internet of Things (IIoT) powered by blockchain technology may successfully fend off dangers of data loss and malicious manipulation brought on by hostile attacks on any one node device (Bhambri et al., 2023).

Data-Sharing: Blockchain now provides for the simultaneous recording and exchange of information by all participants, thanks to the adoption of DLT. This facilitates the sharing of real-time data between upstream and downstream businesses. Confidential information is encrypted in order to resolve the conflicts between data privacy and data-sharing at the same time since it has a privacy protection mechanism. Blockchain technology can end informational silos and enable efficient data-sharing between many stakeholders (Kataria et al., 2023). In a real-world setting, a blockchain-based network for sharing logistics data can expedite document transmission and account reconciliation, increasing logistics efficiency overall (Rani et al., 2023e).

Trust Mechanism: A "trust machine," or blockchain technology, can offer trust services to all partners in smart manufacturing. The crucial data used in the design, production, and selling processes is jointly kept using reliable techniques. Transparent and reliable information is shared by producers, distributors, suppliers, and other intelligent manufacturing entities. As a result, trusting relationships are developed between various parties. The use of blockchain technology in practical settings can eliminate supplier background checks and product quality inspection, bringing the cost of smart manufacturing down even more (Singh & Rani, 2023).

System Coordination: The efficiency of transactions for businesses is considerably increased by paperless automatic transactions and reliable electronic storage based on blockchain. Additionally, the smart contract of blockchain technology encourages cooperation between upstream and downstream businesses in the supply chain of smart manufacturing. It thus becomes a powerful force for the coordinated growth of the entire smart manufacturing system. In a practical setting, a blockchain-based smart procurement platform can encourage procurement coordination and enhance the transparency of trade ties within the sector.

1.4 INTEGRATION OF INTERNET OF THINGS (IoT) AND BLOCKCHAIN TECHNOLOGY

By boosting security, transparency, and efficiency, blockchain technology and the Internet of Things (IoT) are two breakthrough technologies that have the potential to completely change a variety of sectors. When combined, they produce an effective synergy that can deal with many of the problems brought on by IoT. Blockchain is a decentralized and distributed ledger system that was first created as the foundational technology for cryptocurrencies like Bitcoin. It secures and immutably records transactions, making it challenging for a single party to change the data. A transparent and continuous chain of data is produced because each block in the chain carries a cryptographic reference to the block before it as the "Internet of Things." These gadgets might include everything from wearable fitness trackers to smart municipal infrastructure to domestic appliances and commercial machines. Data security, privacy challenges, and the requirement for confidence in the data produced by these devices are just a few of the major problems the IoT must deal with (Rani et al., 2023a).

IoT and blockchain integration provide the following benefits:

Enhanced Security: Blockchain gives IoT data storage access to a safe and impenetrable ledger. Data integrity is ensured by recording each transaction or piece of data in a way that prevents modification or deletion. For sensitive applications like healthcare and industrial control systems, this is essential.

Data Security: Using cryptographic keys, blockchain enables users to manage who has access to their IoT data. Users may control who has access to their data and provide authorization for others to do so, improving privacy and ensuring compliance with data protection laws like the GDPR.

Enhanced Trust: The blockchain eliminates the need for middlemen or centralized bodies to verify IoT data. Data validity is guaranteed by this decentralized trust paradigm, lowering the possibility of fraud or manipulation.

Efficient Transactions: Blockchain can automate microtransactions within IoT networks, allowing devices to execute safe transactions on their own. Smart appliances, for example, may purchase electricity from the grid and settle bills automatically.

Supply Chain Transparency: In the field of supply chain management, blockchain technology can provide complete item visibility and traceability. The blockchain may retain a unique digital identity for each item, reducing fraud and guaranteeing product authenticity.

1.5 FUTURE DEVELOPMENTS OF BLOCKCHAIN TECHNOLOGY: EMERGING TRENDS

Blockchain technology has developed much beyond its first use, such as the supporting infrastructure for cryptocurrencies like Bitcoin. Blockchain is experiencing a number of new trends and future advancements as it continues to evolve, which have the potential to completely alter industries and the digital world. We examine these developments in depth in this thorough investigation, talking about their possible effects and the creative initiatives advancing them.

1.5.1 Quantum Computing and Security for Blockchain

The imminent danger posed by quantum computing is one of the most important new trends in the blockchain industry. With their tenfold increased processing capability, quantum computers have the potential to defeat the cryptographic techniques that underpin blockchain security. This prompts questions about the blockchain networks' long-term security. To lessen this threat, researchers and developers are working hard to build quantum-resistant cryptographic techniques. Blockchain systems are being protected from quantum assaults using post-quantum cryptography, which includes lattice-based encryption and hash-based cryptography. Especially noteworthy are initiatives like NIST's Post-Quantum Cryptography Standardisation that aim to develop cryptographic standards resistant to quantum assaults.

1.5.2 Integration and Interoperability across Chains

As many blockchain networks have evolved, each with its own distinct features and capabilities, interoperability has become a crucial area of study. The necessity for cross-chain integration has sparked the creation of tools and initiatives that try to streamline communication among various blockchain networks. For instance, Polkadot uses a multi-chain architecture to link and securely communicate information between several blockchains. This makes it possible to move assets and data between chains, creating a network of interconnected blockchains. Cross-chain protocols like Wrapped Bitcoin (WBTC) show the promise of interoperable blockchain systems by bringing Bitcoin's liquidity to the Ethereum network.

1.5.3 Financial Decentralization (DeFi)

Unquestionably, one of the most revolutionary aspects in the blockchain industry is decentralized finance, or DeFi. Decentralized financial intermediaries (DeFi) are a group of financial services and apps based on blockchain technology that aspire to take the role of centralized financial intermediaries. Without the need of banks or middlemen, users may lend, borrow, and earn interest on cryptocurrencies, thanks to projects like Compound, Aave, and MakerDAO. The amount of money locked up in DeFi protocols has risen into the billions, showing that the market is becoming more widely accepted and trusted. DeFi offers innovative ways to expand financial inclusion, but it also poses issues with security, government oversight, and the possibility for smart contract weaknesses.

1.5.4 Artificial Intelligence (AI) and Blockchain

Another new development with significant ramifications is the fusion of blockchain with artificial intelligence (AI). Blockchain technology offers an unchanging and transparent record of data and model training, which can increase the reliability of

AI systems. Blockchain networks enable the safe storage and exchange of AI models, lowering the possibility of data tampering and guaranteeing model provenance. Users will be able to purchase and sell AI services and data safely, thanks to projects like Ocean Protocol and SingularityNET, which aim to build decentralized AI markets. The combination of blockchain technology and AI is anticipated to hasten the adoption of AI in a variety of sectors, including banking and healthcare, while resolving issues with data privacy and algorithmic trust.

1.5.5 Sustainable Blockchain Technology

Sustainability is a factor that is being given more weight in the blockchain industry. Many blockchain networks use energy-intensive consensus algorithms like Proof of Work (PoW), which have been criticized for their negative effects on the environment. As a result, there is a rising movement to embrace consensus algorithms that are more environment-friendly. With Ethereum 2.0, for instance, Proof of Stake (PoS) is replacing Proof of Work (PoW), greatly reducing energy usage. Additionally, renewable energy and carbon credits are emphasized in blockchain initiatives like Energy Web and Power Ledger, encouraging sustainability in the energy sector. This transition to green blockchains is part of the blockchain industry's greater commitment to environmental responsibility.

1.6 REAL-WORLD CASE STUDIES

1.6.1 Transforming Food Supply Chains with IBM Food Trust

Case Study: IBM Food Trust is a blockchain-based platform that attempts to improve the supply chain's transparency and traceability. On an unchangeable blockchain ledger, parties may track a food item's full path from farm to table. Major businesses in the food sector, like Walmart and Nestlé, have used the platform.

Details: Consumers and merchants may track the origin and travel of food goods with the help of the IBM Food Trust, ensuring their quality and authenticity. This openness makes it easier to spot and handle food safety problems promptly. For instance, in 2018, when a batch of romaine lettuce was found to be infected, Walmart utilized IBM Food Trust to quickly identify the source of the infection, as opposed to taking days or even weeks using conventional techniques. This skill is essential for stopping foodborne infections and keeping customer confidence.

1.6.2 Everledger: Safeguarding the Diamond Sector

Case Study: The blockchain-based platform Everledger was created to trace diamond origins and stop the trade in conflict diamonds. Each diamond's origin, certification, and ownership history are all carefully documented.

Details: The blockchain system from Everledger creates a tamper-proof database that guarantees the validity of diamonds across the supply chain (Rani et al., 2023c). It aids in the fight against the trade in conflict diamonds by confirming the authenticity of diamonds and their moral origins (Rani et al., 2023c). This supports ethical sourcing and fair-trade practices in addition to protecting the diamond industry's reputation.

1.6.3 Enabling Peer-to-Peer Energy Trading with Power Ledger

Case Study: The blockchain-based platform Power Ledger makes peer-to-peer energy trade possible. It enables direct exchange of surplus renewable energy between people and companies.

Blockchain technology from Power Ledger enables safe and open energy transactions. For instance, in Australia, people who have solar panels may use a platform to sell excess energy to their neighbors, decreasing their dependency on centralized energy suppliers and encouraging the use of renewable energy sources. A more decentralized and sustainable energy ecology is supported by this invention.

1.6.4 Transforming Cross-Border Payments with Ripple

Case Study: Ripple is a blockchain-based technology created to simplify international transfers and payments. It strives to eliminate the lag times and inefficiencies present in conventional financial systems.

Information: Ripple's blockchain technology and digital currency, XRP, make cross-border transactions almost instantly and affordably possible. Worldwide financial institutions have embraced Ripple's solutions to increase the efficiency and dependability of cross-border money transactions. For instance, Santander, a significant international bank, introduced its One Pay FX platform, backed by Ripple, to provide its clients with quicker and more transparent cross-border payments.

1.6.5 Myco: Using Blockchain to Ensure Ethical Fashion

Case Study: Myco is a blockchain-based platform created to advance moral and environment-friendly behavior in the garment sector. Customers can follow the path that apparel takes from manufacture to purchase, thanks to this (Singh and Rani, 2023).

Details: The fashion supply chain benefits from openness and accountability, thanks to Myco's blockchain solution. Consumers may get details on the components, manufacturing processes, and environmental effects of a garment by scanning a QR code on the label. Myco encourages customers to make more knowledgeable and moral purchase decisions, which helps to create a more sustainable fashion sector.

These additional case studies demonstrate how blockchain technology is used to tackle various problems, such as energy distribution, security, and moral consumer decisions in the fashion sector. They serve as an example of how blockchain may be used to improve efficiency, trust, and transparency in a variety of industries.

1.7 CONCLUSION

To sum up, blockchain technology is capable of bringing a paradigm shift in a wide range of industries and enterprises in the future. Blockchain's decentralized and open architecture may improve security, boost productivity, and foster trust across a wide range of applications. The results suggest that through simplifying payment processes, lowering transaction costs, and increasing financial inclusion, blockchain technology can transform the financial sector. By offering a tamper-proof and visible record of transactions, enhancing traceability, and removing fake goods, it may also have a substantial influence on supply chain management. This technology has the potential to enhance data privacy, interoperability, and integrity in the healthcare industry, resulting in improved patient care and results. Blockchain can help governments and public institutions enhance public services and governance by providing safe and transparent voting processes, decreasing corruption, and expediting administrative operations. Furthermore, blockchain has the potential to play an important role in protecting IoT devices, simplifying safe data-sharing, and enabling autonomous machine-to-machine interactions. Despite the many advantages of blockchain technology, there are still issues that need to be resolved. Scalability remains to be an ongoing challenge for blockchain networks as they struggle to handle numerous transactions. Another problem is energy use, since certain networks demand a lot of processing power. Additionally, regulatory frameworks must be created to guarantee security, privacy, and compliance with the law. The findings show that in order to promote innovation, fix problems, and realize the full potential of blockchain technology, industry leaders, governments, and academia must work together. In order to solve problems with scalability and energy consumption, as well as to create strong regulatory frameworks, further research and development is required. To summarize, blockchain technology has a promising future. Its ability to improve security, efficiency, and trust makes it a game changer in a variety of businesses. Blockchain technology can pave the way for a more secure, efficient, and trustworthy future by tackling issues and utilizing collaborative efforts.

REFERENCES

Bali, V., Bali, S., Gaur, D., Rani, S., & Kumar, R. (2023). Commercial-off-the shelf vendor selection: A multi-criteria decision-making approach using intuitionistic fuzzy sets and TOPSIS. *Operational Research in Engineering Sciences: Theory and Applications*, *6*(2).

Bhambri, P., & Rani, S. (2024). Challenges, opportunities, and the future of industrial engineering with IoT and AI. In *Integration of AI-Based Manufacturing and Industrial Engineering Systems with the Internet of Things* (pp. 1–18). USA: CRC Press.

Bhambri, P., Rani, S., Balas, V. E., & Elngar, A. A. (Eds.). (2023). *Integration of AI-Based Manufacturing and Industrial Engineering Systems with the Internet of Things*. USA: CRC Press.

Kataria, A., Puri, V., Pareek, P. K., & Rani, S. (2023, July). Human activity classification using G-XGB. In *2023 IEEE International Conference on Data Science and Network Security (ICDSNS)* (pp. 1–5). Tiptur: IEEE.

Kumar, P., Banerjee, K., Singhal, N., Kumar, A., Rani, S., Kumar, R., & Lavinia, C. A. (2022). Verifiable, secure mobile agent migration in healthcare systems using a polynomial-based threshold secret sharing scheme with a blowfish algorithm. *Sensors*, *22*(22), 8620.

Puri, V., Kataria, A., Solanki, V. K., & Rani, S. (2022, December). AI-based botnet attack classification and detection in IoT devices. In *2022 IEEE International Conference on Machine Learning and Applied Network Technologies (ICMLANT)* (pp. 1–5). Slvador: IEEE.

Rani, S., Bhambri, P., & Kataria, A. (2023a). Integration of IoT, big data, and cloud computing technologies. In *Big Data, Cloud Computing and IoT: Tools and Applications*. USA: CRC.

Rani, S., Kaur, J., & Bhambri, P. (2023b). Technology and gender violence: Victimization model, consequences and measures. In *Communication Technology and Gender Violence* (pp. 1–19). Cham: Springer International Publishing.

Rani, S., Kumar, S., Kataria, A., & Min, H. (2023c). SmartHealth: An intelligent framework to secure IoMT service applications using machine learning. *ICT Express*, *10*(2), 425–430.

Rani, S., Mishra, A. K., Kataria, A., Mallik, S., & Qin, H. (2023d). Machine learning-based optimal crop selection system in smart agriculture. *Scientific Reports*, *13*(1), 15997.

Rani, S., Pareek, P. K., Kaur, J., Chauhan, M., & Bhambri, P. (2023e, February). Quantum machine learning in healthcare: Developments and challenges. In *2023 IEEE International Conference on Integrated Circuits and Communication Systems (ICICACS)* (pp. 1–7). Raichur: IEEE.

Singh, H., & Rani, S. (2023). Flex sensor integrated smart strap to verify correct wearing of the face mask. *IEEE Sensors Journal*, *24*(2), 2020–2027.

2 Evolution of IoT in Various Application Domains

V. Sridhar, Sita Rani, Piyush Kumar Pareek, and Pankaj Bhambri

2.1 INTRODUCTION

The network of physically connected "things" that are integrated with sensors, software, and other technologies to exchange data with other devices and systems over the internet is known as the Internet of Things (IoT). These gadgets might be anything from commonplace items like wearable technology and home appliances to sophisticated machinery and infrastructural parts (Rose et al., 2015). The IoT aims to make it possible for these gadgets to gather and exchange data, make deft decisions, and frequently function independently. This will boost productivity, automate processes, and enhance functionality across a range of industries, including smart cities, healthcare, transportation, and agriculture. Major business trends of IoT in 2023 are shown in Figure 2.1.

Across various industries, IoT improves productivity, connection, and ease of use. It makes process optimization possible by enabling automation, predictive maintenance, and real-time data monitoring. It makes remote patient monitoring easier in the healthcare industry. It enhances infrastructure management in smart cities (Koohang et al., 2022). IoT generally has disruptive effects that raise living standards, increase productivity, and lower prices. Smart homes, connected cars, industrial automation, agricultural optimization, and health monitoring are examples of IoT applications. It facilitates smooth data transfer across gadgets, which promotes productivity, automation, and better decision-making.

2.1.1 Historical Background

The concept of linking things to the internet first surfaced in the early 1980s, which is when IoT got its start. The first noteworthy example was the 1982 Coca-Cola machine at Carnegie Mellon University that was connected to the internet. In the 1990s, British technologist Kevin Ashton first proposed the phrase "Internet of Things," imagining a time when common things could interact with each other through the internet. The Internet of Things' growth was further propelled by advancements in sensor networks and RFID technologies. Significant advancements in device shrinking and

DOI: 10.1201/9781003460367-2

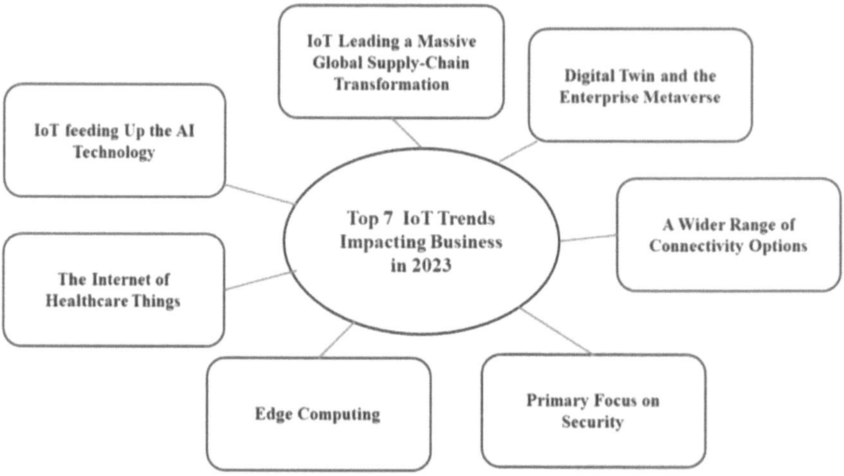

FIGURE 2.1 Major IoT business trends.

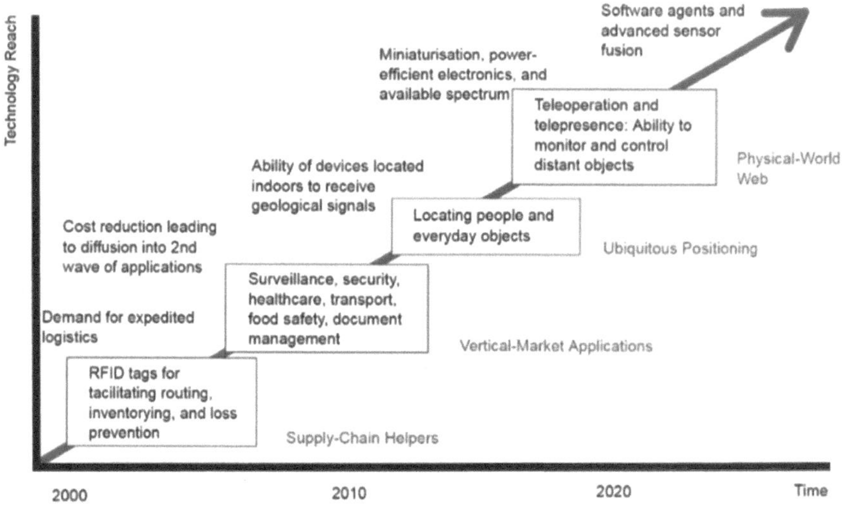

FIGURE 2.2 The historical development of IoT.

connectivity occurred in the late 1990s and early 2000s, setting the stage for the widespread use of IoT.

IoT applications expanded throughout a variety of industries as the twenty-first century went on, including smart cities, healthcare, and agriculture. IoT growth was spurred by developments in data analytics, cloud computing, and wireless communication. IoT is a revolutionary force now, bringing billions of devices together globally, increasing efficiency, automating tasks, and influencing how technology and connection develop in the future. The historical development of IoT is shown in Figure 2.2.

2.2 TECHNOLOGICAL FOUNDATIONS OF IoT

The IoT technology underpinnings comprise a complex ecosystem that facilitates intelligent decision-making, data exchange, and seamless connectivity (Tao et al., 2021). The following are the crucial components:

- *Sensor Technologies:* IoT gathers data from the real world primarily through a variety of sensors. In addition to measuring environmental variables like temperature and humidity, these sensors can also detect movements and track other health measures. Miniaturization and cost reduction advances have made it easier to integrate sensors widely into a wide range of devices.
- *Connectivity Protocols:* IoT uses a number of connectivity protocols to allow devices to communicate with one another. Cellular networks, Bluetooth, Zigbee, Wi-Fi, and RFID are a few common examples. Many times, variables like data transfer speed, power consumption, and range influence the protocol selection.
- *Data Storage and Processing:* A key element of IoT infrastructure is cloud computing, which offers centralized and scalable data processing and storage capabilities. Devices can offload data for analysis via cloud platforms, providing insightful information. This is enhanced by edge computing, which processes data nearer to the source to minimize latency and improve real-time decision-making.
- *Security Measures:* Strong security measures are essential because of the sensitive nature of the data transferred in IoT networks. Data is protected during transmission and storage by permission mechanisms, secure authentication, and encryption. Secure bootstrapping and device identity management are also covered by security considerations.
- *Interoperability Standards:* Interoperability standards are essential to Internet of Things (IoT) because they facilitate smooth communication and cooperation between systems and devices from various manufacturers. Message Queuing Telemetry Transport (MQTT), Constrained Application Protocol (CoAP), and Object Linking and Embedding for Process Control Unified Architecture (OPC UA) are examples of common standards.
- *Artificial Intelligence (AI) and Machine Learning (ML):* The IoT is improved by AI and ML technologies because they allow devices to recognize patterns in data and make wise decisions. Examples of how AI and ML improve the effectiveness and independence of IoT applications include adaptive control systems, anomaly detection, and predictive analytics.

2.3 EVOLUTION OF IoT

The IoT has developed from simple linked gadgets to a large network of intelligent, interconnected systems. Technological developments in sensors, edge computing, and communication protocols have made it possible for devices to integrate seamlessly, revolutionizing industries, increasing productivity, and opening the door to a future that is increasingly automated and networked.

Over several decades, the IoT has undergone a revolutionary journey. The idea first surfaced at Carnegie Mellon University in the early 1980s when a modified Coke machine was connected, signifying the beginnings of an interconnected world. Projects like the European Union's "ISTAG" and MIT's "Things That Think" investigated the incorporation of physical things into the digital domain in the 1990s.

With the introduction of the term "IoT" and the convergence of technologies like RFID, wireless sensor networks, and pervasive connectivity, the 2000s were a watershed year. The RFID system developed by the Auto-ID Center became a prominent prototype that demonstrated the possibility of tracking and controlling objects in the digital realm. In the meantime, the concept of ubiquitous, context-aware computing was presented by initiatives like Ambient Intelligence and Smart Dust.

IoT grew in popularity in the 2010s as a result of the widespread availability of low-cost sensors, more bandwidth, and cloud computing. Wearables, smart homes, and industrial IoT gained popularity as examples of how connected gadgets might be used in real-world situations. Interoperability was made easier by standardization initiatives, such as the creation of communication protocols like MQTT and CoAP.

IoT grew into a pervasive influence in the 2020s, influencing a variety of sectors including healthcare, agriculture, and industry. In order to process data closer to the source, minimize latency, and increase efficiency, edge computing was developed. IoT systems are now even more capable of analyzing large volumes of data and producing insightful analysis, thanks to the incorporation of artificial intelligence. IoT security measure advances are a result of privacy and security concerns.

2.4 APPLICATION DOMAINS

Applications for IoT can be found in many different fields, such as smart cities, industrial automation, smart homes, healthcare, and agriculture. It improves efficiency, sustainability, and safety in a variety of industries by enabling effective resource management, automation, and real-time monitoring. Various application domains of IoT are shown in Figure 2.3.

2.4.1 Smart Homes

IoT applications improve security, energy efficiency, and convenience in smart homes. Smart appliances, lighting controls, and thermostats are examples of connected devices that optimize energy use through automation and remote control. IoT-integrated smart security systems provide real-time alerting and monitoring. Smart speakers and voice-activated assistants provide smooth communication. IoT-enabled sensors and cameras improve domestic security (Pal et al., 2018). Devices for tracking health are a part of individualized well-being. All things considered, IoT turns conventional houses into responsive, intelligent spaces that give homeowners more control, energy savings, and peace of mind through networked and automated solutions.

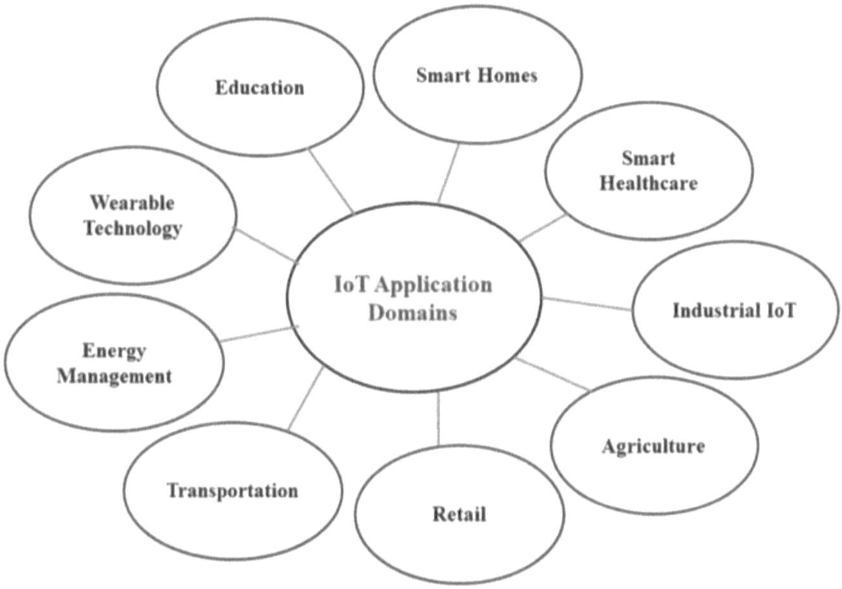

FIGURE 2.3 Various IoT application domains.

2.4.2 Smart Healthcare

IoT is revolutionizing remote diagnostics, medication adherence, and patient monitoring in smart healthcare. Real-time health data is gathered by wearable technology and sensors, allowing for ongoing observation and prompt intervention. IoT integration allows smart medical devices—like insulin pumps—to maximize therapy. Optimizing hospital workflows, maintaining patient information, and managing medical assets more efficiently all help healthcare providers (Rani, Kataria, et al., 2023). By enabling remote consultations, IoT-enabled telemedicine lessens the need for in-person visits. All things considered, IoT applications in smart healthcare improve patient outcomes, expedite the provision of care, and support a more responsive and customized medical ecosystem.

2.4.3 Industrial IoT

IoT applications streamline procedures, boost productivity, and improve safety in industrial environments. IoT sensors and gadgets keep an eye on the condition of the machinery, allowing for predictive maintenance to stop malfunctions. Process improvement is aided by the real-time data collection for performance analytics made possible by Industrial IoT (IIoT) (Boyes et al., 2018). Production workflows are streamlined by automated systems and connected gear, which decreases downtime. IoT-enabled tracking helps supply chain management by guaranteeing smooth transportation. IoT-enabled conditions monitoring for dangers improves safety. All things considered, IoT revolutionizes industry operations by offering useful insights, boosting output, and cultivating a more intelligent and networked manufacturing ecosystem.

2.4.4 Agriculture

IoT applications are revolutionizing farming techniques in agriculture. To optimize irrigation and resource utilization, smart sensors gather data in real time on crop health, temperature, and moisture content of the soil. IoT-guided automated machinery improves precision farming by increasing planting and harvesting efficiency. Drones with IoT devices track vast areas and provide information about agricultural conditions (Rani, Mishra, et al., 2023). IoT-enabled tracking helps livestock management by guaranteeing optimal health and productive breeding. IoT device data analytics support disease diagnosis, crop planning, and decision-making. All things considered, IoT makes agriculture a more data-driven, networked sector that supports sustainability, higher yields, and efficient resource management.

2.4.5 Retail

IoT applications in retail revolutionize operational efficiency and customer experiences. RFID tags and smart shelves simplify inventory management, cutting down on overstock and stockouts. By enabling location-based promotions, beacons and sensors improve tailored marketing. Point-of-sale systems with IoT connectivity enhance consumer relations and transaction efficiency. Shipment tracking in real time improves supply chain visibility. IoT is used in smart retail locations for security, lighting, and climate management (Caro & Sadr, 2019). Fitting rooms and smart mirrors increase customer engagement. Retailers can improve customer satisfaction, make well-informed decisions, and design seamless, technologically advanced shopping environments with the help of data analytics from IoT devices. In general, IoT changes the retail environment by connecting the digital and physical spheres to provide a more effective and customized purchasing experience.

2.4.6 Transportation

IoT applications transform connection, safety, and efficiency in the transportation sector. IoT sensors are used by smart traffic management systems to optimize traffic flow and lessen congestion. Road safety is improved via vehicle-to-vehicle communication, which offers accident avoidance and real-time alerts. GPS tracking and IoT devices help fleet management by improving fuel usage and routes. Real-time tracking and scheduling improve the responsiveness of public transit (Muthuramalingam et al., 2019). Smart parking systems with IoT capabilities direct cars to open spots, easing traffic. Shipment tracking in real time helps enhance supply chain management and logistics. All things considered, IoT makes transportation a more data-driven, networked environment that improves efficiency, safety, and the mobility experience overall.

2.4.7 Energy Management

IoT applications optimize resource usage, efficiency, and sustainability in energy management. Smart grids use the IoT to monitor and control the distribution of electricity in real time, improving efficiency and cutting down on waste. Buildings equipped with IoT sensors can use smart energy management by modifying HVAC and lighting settings in response to occupancy and outside factors. Energy monitoring systems keep tabs on consumption trends, which helps with well-informed conservation decision-making. Renewable energy infrastructure, like solar panels and wind turbines, benefits from IoT-enabled devices that improve upkeep and performance (Wei et al., 2016). In general, real-time insights, proactive conservation measures, and the development of a more robust and sustainable energy ecosystem are how IoT advances energy management.

2.4.8 Wearable Technology

Wearable technology uses the IoT to improve connectivity and individual well-being. IoT sensors are used by fitness trackers and smartwatches to detect activity, heart rate, and sleep habits, hence raising awareness of health issues. IoT-enabled smart apparel gives athletes access to real-time biometric data and healthcare tracking. IoT is used by wearable medical equipment to enable remote patient monitoring, such as continuous glucose monitors (Rani, Kumar, et al., 2023; Singh & Rani, 2023). A unified and customized user experience is produced by these devices' seamless connections to smartphones and other smart devices. In general, wearable technology IoT applications support a more linked, health-conscious, and digitally savvy lifestyle.

2.4.9 Education

IoT uses in education improve administrative effectiveness and learning outcomes. IoT devices are integrated into smart classrooms to facilitate interactive learning, and connected whiteboards and devices increase student participation. Wearables and sensors with IoT capabilities keep an eye on students' attendance and well-being. Inventory management is streamlined by asset tracking systems. Personalized learning is made possible by educational platforms that are connected to the cloud. IoT is used by smart campuses for building efficiency, security, and energy management. Virtual classrooms made possible by IoT-enabled gadgets are beneficial for remote education (Al-Emran et al., 2020). All things considered, IoT revolutionizes education by encouraging creativity, enhancing administrative procedures, and building a more dynamic and interconnected educational ecosystem.

2.5 CHALLENGES AND CONSIDERATIONS

The main obstacles facing IoT applications are security flaws, privacy issues, interoperability problems, and the shear complexity of managing a variety of devices. Other challenges include resolving data governance and guaranteeing dependable connectivity.

2.5.1 Security and Privacy Concerns

In IoT applications, security and privacy issues are major problems. Unauthorized access and data breaches can result from vulnerabilities in connected devices. Information is sensitive to compromise due to inadequate authentication and encryption procedures. The ecosystem is vulnerable to cyber threats because of the sheer number of linked devices, which expand the attack surface (Hameed & Alomary, 2019). The widespread collection and sharing of personal data puts privacy at danger as well. It is difficult to strike a balance between protecting user privacy and using data for usefulness. To address these issues and safeguard users from potential dangers, reliable security mechanisms, frequent upgrades, and industry-wide standards are needed to guarantee the reliability of IoT apps.

2.5.2 Interoperability Issues

IoT applications continue to face interoperability issues because of the variety of devices and communication protocols. Compatibility issues prevent devices from different manufacturers from exchanging data seamlessly, which impedes the growth of cohesive ecosystems. In order to provide uniform data formats and communication protocols, standardization activities are crucial. IoT solutions' scalability and efficiency are constrained by the complexity of integrating devices into unified systems in the absence of compatibility (Noura et al., 2019). In order to overcome these obstacles and promote a more interconnected and interoperable IoT landscape that can fully realize the potential of a cohesive and integrated smart environment, industry collaboration, open standards, and attempts to develop common frameworks are essential.

2.5.3 Scalability Challenges

In IoT applications, scalability becomes a major difficulty as the number of connected devices increases quickly. Infrastructure and analytics capabilities are put to the test when managing and analyzing the massive volumes of data produced by a growing ecosystem. Obstacles include greater energy usage, latency problems, and network congestion. It becomes complex to maintain a constant quality of service across a large-scale deployment, which affects responsiveness and efficiency (Luntovskyy & Globa, 2019). To meet the scalability requirements of the IoT, solutions must have strong architectures, edge computing methodologies, and optimized communication protocols. To fully realize the potential of IoT applications, it is imperative to strike a balance between the expansion of connected devices and scalable infrastructure and data management systems.

2.6 FUTURE TRENDS AND INNOVATIONS

IoT future trends include edge computing for quicker processing, expanded connectivity via 6G networks, integration of AI for more intelligent analytics, and a greater emphasis on IoT security. More intelligent and efficient applications will be driven by innovations in a variety of industries.

2.6.1 Edge Computing Advancements

Increased use of real-time data processing, the incorporation of AI for more intelligent edge decision-making, improved security protocols, and the growth of decentralized edge ecosystems are some of the future trends for IoT edge computing. As edge computing spreads, it will spur advancements in effective data management, lower latency, and facilitate the expansion of IoT applications in a variety of sectors.

2.6.2 Integration with Artificial Intelligence

Future IoT and AI integration will see a sharp increase in the use of Edge AI for real-time processing, which will lessen dependency on centralized servers. Explainable AI will become more popular, guaranteeing decision-making that is transparent and trustworthy. Operations will be streamlined by AI-driven automation, and proactive insights from IoT data will be obtained through predictive analytics. AI will strengthen IoT ecosystems against cyberattacks with improved security mechanisms. Standards for interoperability will advance to enable smooth cooperation, creating an AIoT environment that is more linked and effective. This integration points to a future in which intelligent, data-driven results are produced by smart, AI-powered IoT apps that optimize processes and change industries.

2.6.3 Emerging IoT Applications

The broad use of IoT in healthcare for remote patient monitoring, the incorporation of IoT in smart cities for increased sustainability and efficiency, and the emergence of IoT in agriculture for precision farming are some future trends in emerging IoT applications. IoT applications for environmental monitoring are anticipated to expand, addressing concerns like air quality and climate change mitigation. Industries will also use IoT more and more for supply chain optimization.

2.7 CONCLUSION

The IoT development has transformed many fields. It facilitates individualized care and remote monitoring in the medical field. IoT is used by smart homes to automate tasks and save energy. Streamlined operations and predictive maintenance are beneficial to industries. IoT is used in transportation to improve efficiency and safety. IoT is used in energy management to maximize resource use. IoT is used by retail to manage inventory and provide individualized consumer experiences. IoT has brought about smart classrooms and increased administrative efficiency in education. Precision farming in agriculture is being embraced by IoT-enabled solutions. IoT is used by wearable technologies to check health. The ongoing development of IoT in these disciplines is indicative of its revolutionary influence, improving productivity, reach, and user experiences in numerous industries.

Data breaches, unauthorized access risks, and security vulnerabilities are major issues in IoT application fields. The vast amount of data being collected gives rise to privacy issues. Problems with interoperability impede smooth communication between various devices. Infrastructure is being strained by the exponential development of connected devices, which presents scalability difficulties. Additionally, standardization efforts are necessary since maintaining various data formats presents challenges. Reliability, security, and scalability of IoT applications in the transportation, smart home, healthcare, and other sectors depend on addressing these issues.

REFERENCES

Al-Emran, M., Malik, S. I., & Al-Kabi, M. N. (2020). A Survey of Internet of Things (IoT) in Education: Opportunities and Challenges. In *Toward Social Internet of Things (SIoT): Enabling Technologies, Architectures and Applications: Emerging Technologies for Connected and Smart Social Objects*, 197–209, Springer, New York, NY.

Boyes, H., Hallaq, B., Cunningham, J., & Watson, T. (2018). The Industrial Internet of Things (IIoT): An Analysis Framework. *Computers in Industry,* 101, 1–12.

Caro, F., & Sadr, R. (2019). The Internet of Things (IoT) in Retail: Bridging Supply and Demand. *Business Horizons,* 62(1), 47–54.

Hameed, A., & Alomary, A. (2019). *Security Issues in IoT: A Survey*. Paper Presented at the 2019 International Conference on Innovation and Intelligence for Informatics, Computing, and Technologies (3ICT).

Koohang, A., Sargent, C. S., Nord, J. H., & Paliszkiewicz, J. (2022). Internet of Things (IoT): From Awareness to Continued Use. *International Journal of Information Management,* 62, 102442.

Luntovskyy, A., & Globa, L. (2019). *Performance, Reliability and Scalability for IoT*. Paper Presented at the 2019 International Conference on Information and Digital Technologies (IDT).

Muthuramalingam, S., Bharathi, A., Rakesh Kumar, S., Gayathri, N., Sathiyaraj, R., & Balamurugan, B. (2019). IoT Based Intelligent Transportation System (IoT-ITS) for Global Perspective: A Case Study. In *Internet of Things and Big Data Analytics for Smart Generation*, 279–300, Springer, New York, NY.

Noura, M., Atiquzzaman, M., & Gaedke, M. (2019). Interoperability in Internet of Things: Taxonomies and Open Challenges. *Mobile Networks and Applications,* 24, 796–809.

Pal, D., Funilkul, S., Charoenkitkarn, N., & Kanthamanon, P. (2018). Internet-of-Things and Smart Homes for Elderly Healthcare: An End User Perspective. *IEEE Access,* 6, 10483–10496.

Rani, S., Kataria, A., Kumar, S., & Tiwari, P. (2023). Federated Learning for Secure IoMT-Applications in Smart Healthcare Systems: A Comprehensive Review. *Knowledge-Based Systems*, 110658.

Rani, S., Kumar, S., Kataria, A., & Min, H. (2023). SmartHealth: An Intelligent Framework to Secure IoMT Service Applications using Machine Learning. *ICT Express*, 10(2), 425–430.

Rani, S., Mishra, A. K., Kataria, A., Mallik, S., & Qin, H. (2023). Machine Learning-Based Optimal Crop Selection System in Smart Agriculture. *Scientific Reports,* 13(1), 15997.

Rose, K., Eldridge, S., & Chapin, L. (2015). The Internet of Things: An Overview. *The Internet Society (ISOC),* 80, 1–50.

Singh, H., & Rani, S. (2023). Flex Sensor Integrated Smart Strap to Verify Correct Wearing of the Face Mask. *IEEE Sensors Journal*, 24(2), 2020–2027.

Tao, W., Zhao, L., Wang, G., & Liang, R. (2021). Review of the Internet of Things Communication Technologies in Smart Agriculture and Challenges. *Computers and Electronics in Agriculture,* 189, 106352.

Wei, M., Hong, S. H., & Alam, M. (2016). An IoT-Based Energy-Management Platform for Industrial Facilities. *Applied Energy,* 164, 607–619.

3 Transforming Healthcare Crowdfunding

Integrating Blockchain and IoT for Transparency, Trust, and Accessibility

Vani Vasudevan, Udhayaranjani Sellappagounder Mohan, and Mohan S. G.

3.1 INTRODUCTION

The activity of soliciting funds for medical bills, therapies, or healthcare-related projects through internet platforms, social media, or crowdfunding websites is known as healthcare crowdfunding. Individuals, patients, or their families seek financial assistance from a large network of donors, friends, family members, and even strangers to cover the costs of medical treatments, operations, prescriptions, medical equipment, or other healthcare requirements (Rani et al., 2023b). This strategy makes use of the internet and social networks to reach a large audience and collect financial contributions, frequently in the form of small donations, to help reduce the financial burden connected with healthcare bills (Howe, 2006). Healthcare crowdfunding has filled gaps in healthcare coverage, assisting individuals experiencing unforeseen medical issues and funding medical research and creative healthcare ventures which has in turn increased its popularity. Figure 3.1 presents the consistent upward trajectory in publications pertaining to "crowdfunding" from 2013 to 2023, as of September 19, 2023. Noticeable linearity is shown in this progression. It is worth noting that crowdfunding reached an initial peak with 20,900 publications during the COVID-19 pandemic. Figure 3.2 showcases a similar positive trend in publications related to both "crowdfunding" and "healthcare." Figure 3.3 illustrates the publication trend which contains keywords such as "crowdfunding," "healthcare," and "blockchain," while Figure 3.4 illustrates the trend involving the keywords "crowdfunding," "healthcare," and "IoT."

Lastly, Figure 3.5 reveals a gradual yet consistent increase in publications encompassing "crowdfunding," "healthcare," "blockchain," and "IoT" since 2017. With the growing number of publications, it is evident that this area has emerged as a focal point. Consequently, this chapter proposes to delve into this dynamic field,

DOI: 10.1201/9781003460367-3

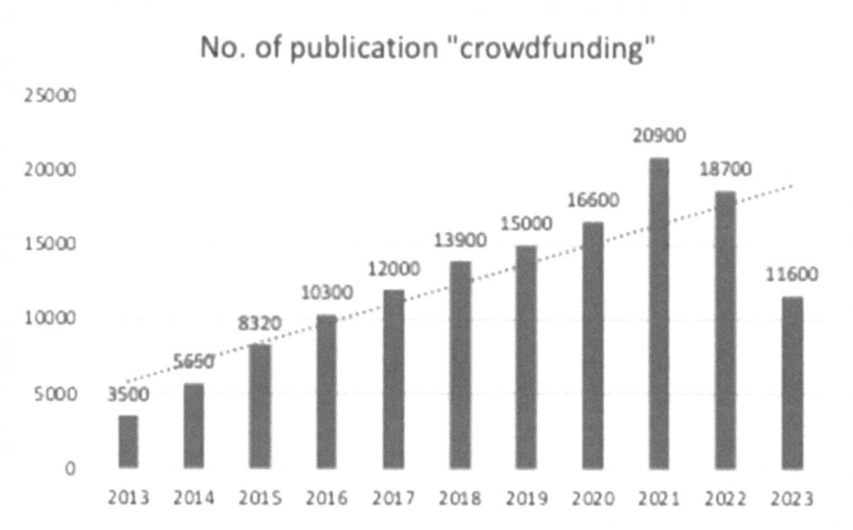

FIGURE 3.1 Number of publications in crowdfunding.

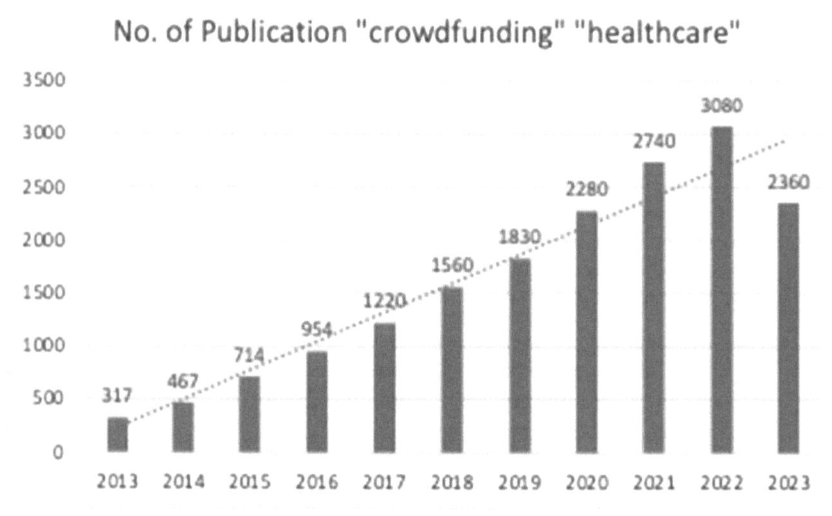

FIGURE 3.2 Number of publications in "crowdfunding" and "healthcare."

highlighting its past developments and shedding light on forthcoming challenges and opportunities.

Healthcare crowdfunding is of considerable importance and has a profound influence on contemporary healthcare for various reasons which include financial assistance for medical expenses (Ghazal et al., 2023), availability of innovative treatments in numerous nations, empowerment and patient-centered care (Kraus et al., 2021),

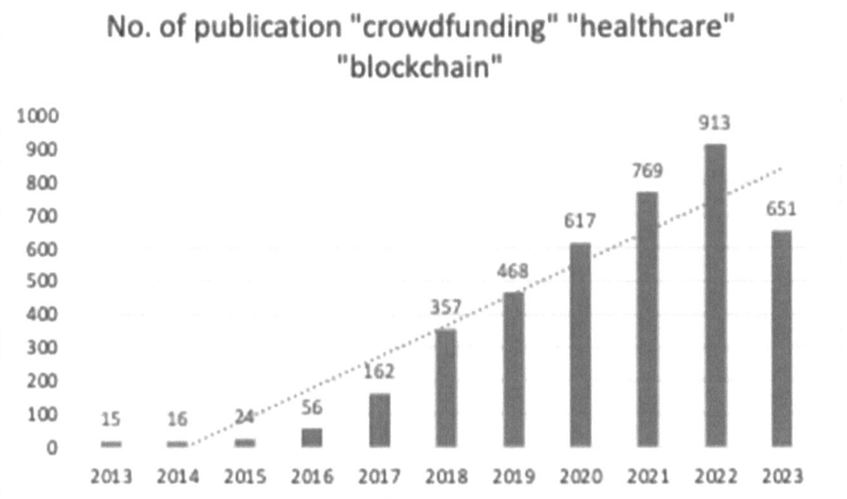

FIGURE 3.3 Number of publications in "crowdfunding," "healthcare," and "blockchain."

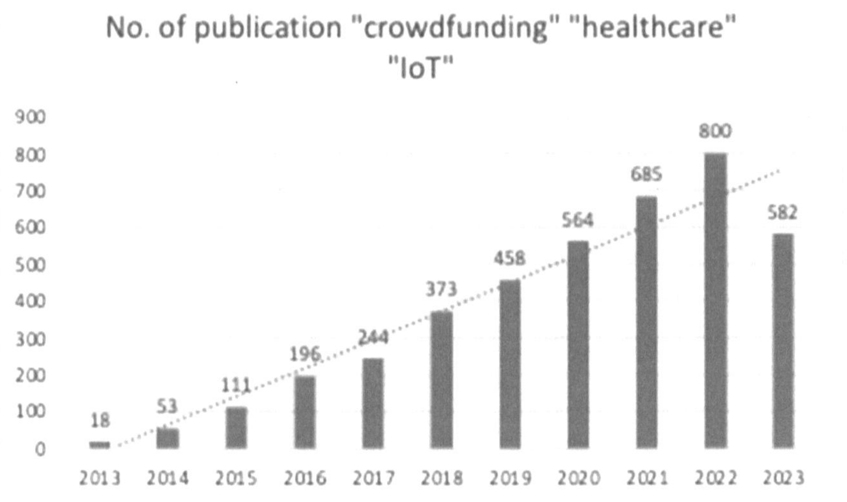

FIGURE 3.4 Number of publications in "crowdfunding," "healthcare," and "IoT."

supporting healthcare research (Bassani et al., 2019; Hou et al., 2022), raising awareness, Community Engagement and Social Assistance (WHO, 2021), data for healthcare trends (Ren et al., 2020), inclusivity, and mitigating healthcare disparities. The expenses associated with healthcare can be excessively high (Ghazal et al., 2023), and not all individuals possess the means to obtain full health insurance coverage. Healthcare crowdfunding serves as a crucial resource for individuals confronted with

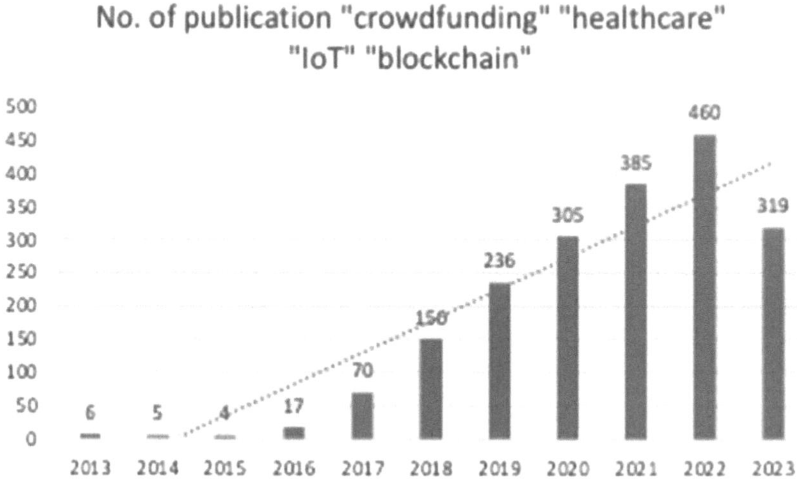

FIGURE 3.5 Number of publications in "crowdfunding," "healthcare," "IoT," and "blockchain" from 2013 to 2023.

unanticipated medical costs (Mollick, 2003) or expenditures that are not included by insurance. This intervention serves to mitigate the financial disparity and guarantee access to essential healthcare services for individuals. Certain medical treatments and therapies now lack widespread accessibility or insurance coverage. Healthcare crowdfunding facilitates patient access to state-of-the-art medicines, experimental therapies, or clinical studies that hold promise and potential remedies for otherwise incurable medical problems.

Patients are empowered by using crowdfunding platforms as it allows them to make informed healthcare choices, thereby fostering patient-centric care (Kraus et al., 2021) approach. Optimal healthcare services and diverse alternative treatments are enabled to individuals as this approach does not limit individuals with financial limitations (Bali et al., 2023). The scope of healthcare crowdfunding surpasses the coverage of individual medical bills. In addition, it has the potential to support medical research, pharmaceutical development, and the innovation of novel healthcare technologies which will improve progress in the medical science field (Rani et al., 2023c). The utilization of crowdfunding campaigns frequently incorporates compelling storylines that serve to increase awareness about particular medical ailments or healthcare concerns. The heightened level of knowledge has the potential to stimulate advocacy endeavors, secure additional money for research, and enhance comprehension of many health-related obstacles. Crowdfunding initiatives frequently garner donations from acquaintances, relatives, and even unfamiliar individuals who exhibit empathy toward the medical circumstances. The presence of a strong community and social support system (WHO, 2021) can yield favorable emotional and psychological outcomes for patients and their families (Kumar et al., 2022).

Healthcare crowdfunding exhibits a high degree of accessibility, since it enables individuals with internet connectivity to engage in fundraising activities, thereby rendering it a very inclusive approach. It surpasses geographical limitations and allows individuals to pursue the support and assistance they require from diverse backgrounds (Sudevan et al., 2021). The mitigation of healthcare disparities can be facilitated by the utilization of healthcare crowdfunding (Sear, 2021) which serves to alleviate financial burdens for persons who face limited access to high-quality healthcare as a result of economic or geographic constraints. India has the largest Facebook audience with 329.65 million users and over 749 million internet users in 2020. Over 1.5 billion users are expected by 2040. India has the second-most smartphone users. All this makes online donation-based crowdfunding in India easier (Puri et al., 2022).

Global crowdfunding is a multibillion-dollar sector, according to World Bank statistics. The research forecasts $300 billion in industry volume by 2030. With the growth of crowdfunding in India, medical fundraising platforms are committed to impacting more lives and communities.

According to researchandmarkets.com report, global crowdfunding market CAGR (Compound Annual Growth Rate) grows at 15.7% in the current year and expected to grow up to 27.81% with the same % of CAGR. According to fundly.com, there are 30+ standout crowdfunding websites for the year 2023. Some of the predominant fundraiser platforms are shown in Table 3.1.

While numerous crowdfunding platforms have emerged over time, conventional healthcare crowdfunding faces several challenges, including transparency and accountability, privacy issues, lack of security, limited accessibility, geographical barriers, regulatory obstacles, administrative expense, data integrity, and reliability as well as fund disbursement issues.

TABLE 3.1
Fundraiser Platforms with Funds Raised for Medical and Healthcare Expenses

Fundraiser Platform	Fund Raised
MyCause: https://www.mycause.com.au/	$10 million+
MyCause is an Australia-based crowdfunding platform that offers both a flexible and a fixed crowdfunding model and let fundraisers choose the campaign	
GoFundMe: https://www.gofundme.com/	$650 million+
GoFundMe is one of the largest and most well-known crowdfunding platforms. While it covers a wide range of causes, it has a significant healthcare and medical expenses category	
FundRazr: https://fundrazr.com/	$230+ million
FundRazr is a crowdfunding platform that allows users to raise funds for various causes, including medical expenses and healthcare treatments	
Watsi: https://watsi.org/	$3.5 million
Watsi is a nonprofit crowdfunding platform that connects donors with patients around the world in need of medical treatments. It focuses on global healthcare initiatives	

3.2 BLOCKCHAIN TECHNOLOGY IN HEALTHCARE

Blockchain technology is driven by cryptocurrencies such as Bitcoin but has implications beyond as it is a decentralized and distributed ledger system. Transactions are recorded throughout a network of computers, or nodes, forming a chain of blocks that store data in this system. Due to the decentralized nature of the technology and cryptographic procedures, data security, transparency and cryptographic procedures are assured. Satoshi Nakamoto's foundational Bitcoin whitepaper (Nakamoto, 2008) established the concept of blockchain and its implementation in a peer-to-peer electronic payment system. Proof of Work (PoW) and Proof of Stake (PoS) consensus mechanisms validate and add new blocks to the chain, while cryptographic hashing and digital signatures secure the information within blocks. A wide range of use cases beyond cryptocurrencies, such as supply chain management, healthcare, and banking, has opened up in consequence of the blockchain data being transparent and accessible to all network participants (Rani et al., 2023a).

Even in sensitive sectors like healthcare, blockchain technology holds great promise as its distributed and secure ledger could eliminate pharmaceutical counterfeiting and ensure patients receive appropriate dosage without any risk of fraud (Dal Mas et al., 2020). Moreover, this technology establishes trust, traceability, and transparency, which brings generous benefits to healthcare. An integrated electronic health record ecosystem with clinical, organizational, and managerial outcome is fostered which is a vital factor in societal sustainability (Rani et al., 2022).

Enhancing data security, interoperability, and trust among stakeholders in the healthcare ecosystem are primary advantages offered by blockchain in healthcare. Patient care, data management, and medical research will be positively impacted by the transparency and immutability of the blockchain records (Rani et al., 2023e). For instance, patient records stored on a blockchain can be accessed and updated securely by authorized healthcare providers, ensuring data accuracy and reducing the risk of errors (Kuo et al., 2017). Interoperability and streamlining in healthcare is made possible as blockchain securely shared health data across institutions (Zhang et al., 2016). It also facilitates clinical trials and medical research by guaranteeing data integrity and enabling secure, consent-based sharing of sensitive patient information.

Integrating blockchain technology in healthcare industry will improve patient care and healthcare innovation and revolutionize the industry by addressing challenges such as data privacy. Figure 3.6 shows various benefits of using blockchain technology in healthcare crowdfunding.

3.3 IoT IN HEALTHCARE

Patient care, data collecting and overall healthcare system is altered efficiently with the disruptive force of Internet of Things (IoT). The IoT allows real-time data gathering, exchange, and analysis as it is a linked network of physical devices, sensors, and software applications. IoT devices in healthcare cover a wide range of applications, from wearable fitness trackers and remote patient monitoring devices to hospital-connected medical equipment. These devices continuously capture and

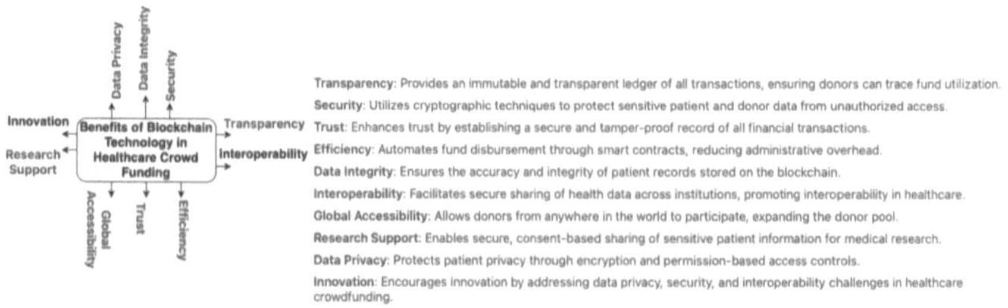

FIGURE 3.6 Benefits of blockchain in healthcare crowdfunding.

FIGURE 3.7 Benefits of IoT in healthcare crowdfunding.

send crucial health data, enabling healthcare providers to remotely monitor patients, make educated clinical decisions, and intervene as needed (Sivarajah et al., 2017). IoT in healthcare can improve patient outcomes, reduce readmissions in hospitals, and improve overall healthcare by offering timely and data-driven insights.

Some notable examples of IoT devices in healthcare contexts include wearable health monitors like Fitbit and Apple Watch, which continuously track physical activity, heart rate, and sleep patterns (Fitbit, n.d.). Continuous glucose monitors (CGMs) such as the Dexcom G6 provide real-time glucose level data for the management of diabetes, allowing individuals to make informed treatment decisions (Dexcom, n.d.). IoT-connected thermometers, such as Kinsa's Smart Thermometer, provide remote temperature surveillance and can notify healthcare providers or caregivers of a fever spike (Kinsa, n.d.). In addition, smart inhalers, such as Propeller Health's inhaler sensors, assist individuals with asthma or chronic obstructive pulmonary disease (COPD) in effectively managing their conditions by monitoring inhaler usage and providing information about their respiratory health (Propeller Health, n.d.).

Real-time patient monitoring, early intervention, and improved medication adherence are offered by IoT devices. Figure 3.7 shows the benefits of IoT in healthcare crowdfunding.

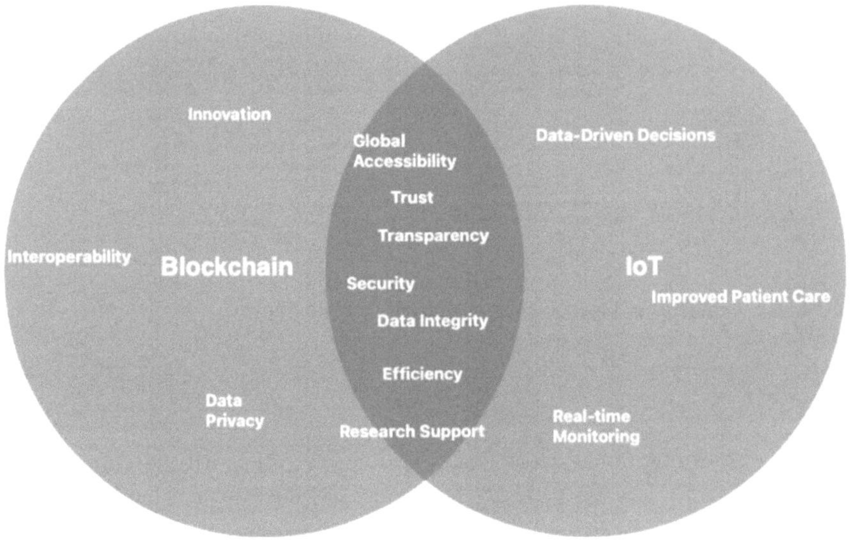

FIGURE 3.8 Benefits of integrated blockchain and IoT in healthcare crowdfunding.

3.4 INTEGRATION OF BLOCKCHAIN AND IoT IN HEALTHCARE CROWDFUNDING

Substantial advancements in transparency, security, and operational effectiveness are revolutionized because of the integration of blockchain technology with IoT in the healthcare crowdfunding industry. Figure 3.8 presents the benefits of this integration.

This chapter explores the potential synergies that might be achieved via the integration of various advanced technologies (Chauhan & Rani, 2021). Figure 3.9 presents the architecture of integrated healthcare crowdfunding system. The use of blockchain technology offers a reliable and unalterable storage system for health data produced by IoT sensors (Li et al., 2017). This includes the documentation of essential patient metrics, adherence to medicine, and progress in treatment, all stored on the blockchain which guarantees the integrity and transparency of data (Rani et al., 2023a). This ensures that confidence is of utmost importance to develop trust among the parties involved in healthcare crowdfunding initiatives (Kataria et al., 2022).

IoT devices are well recognized for their ability to gather and analyze confidential health information. The blockchain technology ensures the protection of patient privacy by securely storing their data (Kshetri, 2017). The use of encryption and permission-based access restrictions inside blockchain technology substantially reduces the potential for unwanted access or data breaches. The cultivation of trust and confidence in crowdfunding projects is facilitated by granting more data ownership to both supporters and patients (Bhambri et al., 2023).

The use of crowdfunding platforms based on blockchain technology surpasses regional boundaries, hence improving worldwide accessibility and accountability (Tama et al., 2017).

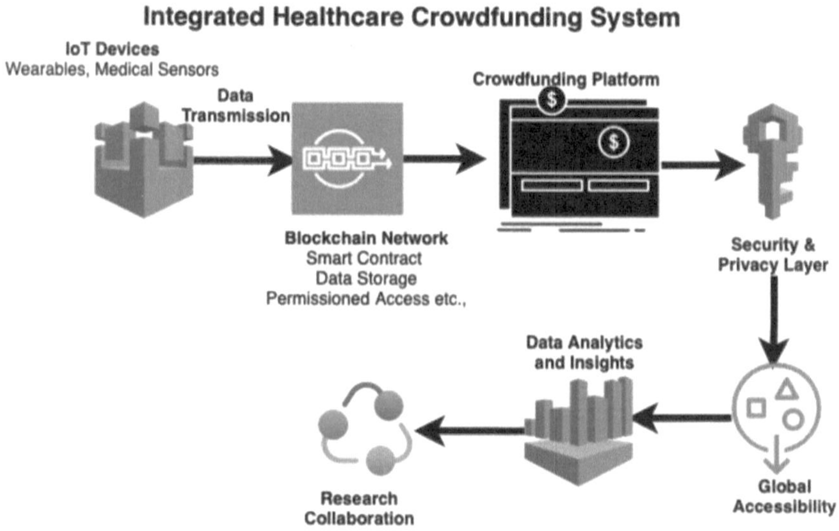

FIGURE 3.9 Architecture of integrated healthcare crowdfunding system.

Promising opportunities for the advancement of data-centric medical research crowdfunding have been introduced by the convergence of blockchain technology and the IoT within the healthcare sector. Those interested in providing financial support for research initiatives find this development in medical knowledge appealing.

The potential advantages of the convergence of blockchain and IoT in healthcare crowdfunding are significant (Rani et al., 2023a). However, it is crucial to recognize and tackle particular issues associated with this integration, including regulatory compliance, scalability, and data accuracy (Zhang et al., 2018). Healthcare crowdfunding industry will potentially be transformed as the use of the technologies promote openness, trust, and accessibility. Figure 3.9 illustrates the flow of data and interactions between the key components in an integrated healthcare crowdfunding system and Figure 3.10 illustrates the process of healthcare crowdfunding with blockchain and IoT integration.

3.4.1 Transparency and Trust in Healthcare Crowdfunding

Blockchain technology is critical to achieving transparency across the healthcare crowdfunding industry. This transparency is supported by a plethora of blockchain technologies. In the beginning, blockchain establishes immutable and transparent ledgers, documenting every transaction with an unchangeable timestamp, enabling the permanent visibility of all financial activities in healthcare crowdfunding (Li et al., 2017). Second, the introduction of smart contracts on the blockchain automates the allocation process based on established parameters, removing the need for intermediaries and limiting the risks of misallocation or fraud. Third, susceptibility of transaction records to manipulation or illegal changes is reduced by the

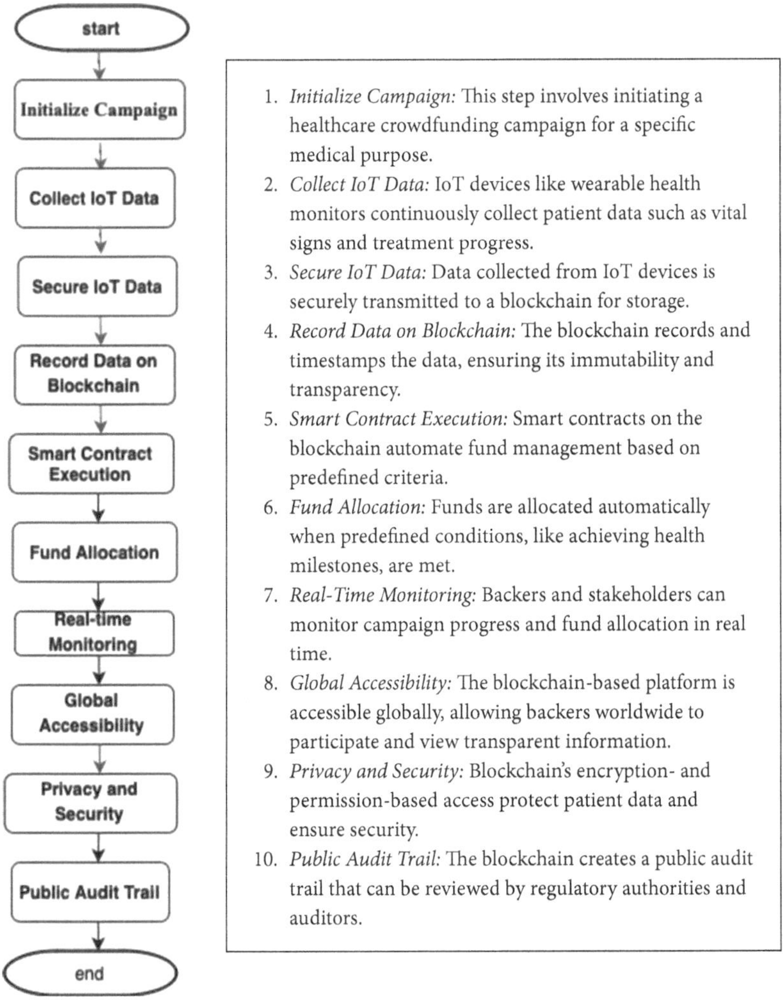

FIGURE 3.10 Process flow of integrated healthcare crowdfunding system.

decentralized network structure of blockchain, reinforcing the principles of transparency and trust (Kshetri, 2017).

Furthermore, the real-time data provided by blockchain on funding distribution and campaign progress allows contributors to assess the impact of their contributions as campaigns advance (Rani et al., 2023d). Wearable health monitors, for instance, are examples of IoT devices that supplement these updates by providing real-time health data and achievement milestones (Tama et al., 2017). The technology also makes it easier to create publicly auditable trails that regulatory agencies and independent auditors may review, increasing accountability and transparency in the allocation process. Finally, by reducing reliance on intermediaries, blockchain reduces the possibility of errors and delays, increasing the transparency of the allocation process (Zhang et al., 2018).

The trust levels and transparency standards of fundraising processes are transformed by the combination of the IoT and blockchain technology in the healthcare crowdfunding arena. Wearable health monitors and medical sensors, for example, offer continuous monitoring and real-time recording of patients' vital signs, treatment progress, and health milestones (Zanella et al., 2014). These continuous data streams are securely logged by blockchain technology, ensuring transparency, immutability, and accessibility (Li et al., 2017). Supporters can track the impact of their contributions by seeing concrete improvements in patient well-being or research progress through continually updated data.

3.4.2 Smart Contracts in Healthcare Crowdfunding

Smart contracts, a fundamental element of blockchain technology, assume a crucial function in the realm of crowdfunding by means of automating and optimizing diverse procedures. Self-executing contracts are characterized by the inclusion of predetermined conditions, which, upon fulfillment, trigger an automatic execution, verification, or enforcement of the contractual provisions (Tanwar et al., 2022). Within the realm of crowdfunding, the implementation of smart contracts serves the purpose of automating the distribution of funds. This mechanism guarantees that the financial contributions made by supporters are only transferred to the project or campaign if predetermined conditions have been met. The utilization of this technology serves the dual purpose of safeguarding investors against potential fraudulent activities and augmenting transparency, as the entirety of the procedure is meticulously documented on an unalterable blockchain ledger (Bhambri & Rani, 2024). Smart contracts have the capacity to decrease administrative burdens, hence enhancing the efficiency of crowdfunding processes (Rani et al., 2023a). Additionally, their inherent resistance to tampering guarantees the integrity of data. Due to the widespread availability of blockchain-based smart contracts, anyone from all around the world have the opportunity to engage as supporters. In addition, the utilization of smart contracts can facilitate the process of tokenization (Mougayar, 2016), which enables supporters to obtain digital tokens that represent their ownership stake in a project. Fractional ownership is promoted effectively and liquidity is enhanced by this mechanism. Within the domain of blockchain technology, smart contracts serve as a pivotal and innovative mechanism for the automated allocation of funds, contingent upon predetermined criteria (Mougayar, 2016; Büyüksaracoğlu et al., 2020). Once the defined requirements have been met, the activities outlined in the contract are executed autonomously. Crucial role is played by smart contracts in ensuring the protection of backers' interests within the realm of crowdfunding. The platform ensures that the financial contributions made by backers are only released to the project or campaign upon the successful attainment of specific milestones or financing objectives. The implementation of automatic payout not only promotes security by reducing the potential for fraudulent campaigns but also greatly improves transparency (Singh & Rani, 2023).

The use of this efficient procedure not only expedites the distribution of funds but also significantly diminishes administrative expenses, resulting in financial benefits (Tapscott et al., 2016). The establishment of trust among stakeholders is facilitated by the transparency that is intrinsic to smart contracts, which are recorded on an immutable blockchain ledger (Mougayar, 2016).

3.4.3 Tokenization and Fractional Ownership

Tokenization, in the realm of blockchain technology and IoT in healthcare crowdfunding, refers to the procedure of transforming tangible assets or entitlements into digital tokens. Blockchain ledger then records these tokens in a secure and transparent manner. Fractional ownership, wherein healthcare assets or projects can be partitioned into smaller, exchangeable parts, is enabled by tokenization. This facilitates increased engagement in crowdfunding campaigns at a more detailed level, even for individuals with limited financial capacity to make substantial individual contributions (Kataria et al., 2023).

According to Büyüksaracoğlu et al. (2020), individuals who wish to engage in the healthcare real estate market have the opportunity to acquire tokens that symbolize partial ownership of these assets. This allows investors to actively participate in this particular sector.

To finance medical research initiatives, tokenization could be utilized. Researchers and organizations have the option to tokenize their projects, which enables potential investors to provide financial support for specific research activities (Tapscott et al., 2016).

3.4.4 Real-Time Data Collection with IoT

A diverse range of IoT devices such as wearable health monitors, medical sensors, and smart medical devices is incorporated in the field of healthcare. Patient monitoring and data collection for medical research and treatment optimization are some of the wide range of applications offered by these devices (Al-Abbasi et al., 2020).

Medical sensors are of utmost importance in the monitoring of certain health indices. For example, the utilization of glucose sensors has been shown to enhance the management of diabetes in patients, since it enables the provision of real-time readings of glucose levels (Prabhakar et al., 2017).

3.4.5 Security and Privacy Considerations

Security concerns must be resolved in the realm of blockchain-based healthcare crowdfunding. Access to unauthorized data and data breaches are the security challenges (Iqbal et al., 2021) foreseen.

Implementing rigorous encryption techniques ensures the confidentiality and security of data stored on the blockchain. Using permissioned blockchain networks restricts data access to only authorized parties. Implementing multi-factor authentication for users who access the blockchain platform adds an additional layer of security. Regularly auditing the blockchain network for vulnerabilities and suspicious activities enables prompt identification and resolution of security concerns.

Retaining the data's utility for analysis and research and removing personally identifiable information will anonymize donor and patient data. Implementing blockchain-based consent mechanisms enables patients to control who has access to their health data, thereby protecting patient privacy. Also, keeping blockchain networks and associated security protocols up-to-date is highly recommended in order to resolve emerging threats and vulnerabilities.

3.4.6 Global Accessibility in Healthcare Crowdfunding

Healthcare crowdfunding will undergo a fundamental shift as a result of the intrinsic potential of blockchain-based platforms to reach a worldwide audience and transcend regional barriers. The global accessibility in healthcare crowdfunding has numerous, significant benefits, which include increasing possibilities for fundraising and various viewpoints and expertise. Crowdfunding campaigns for healthcare can quickly gather money from donors across the world to support pandemic preparedness, disaster relief, or urgent medical needs. Innovative medical ventures and businesses in the healthcare industry can leverage blockchain-based crowdfunding to reach a worldwide audience. Successful global healthcare crowdfunding campaigns include MSF (www.doctorswithoutborders.org) and Patientory (https://patientory.com).

3.5 CHALLENGES AND FUTURE DIRECTIONS

The challenges of regulatory compliance and scalability in the context of integrating blockchain and IoT in healthcare crowdfunding are discussed in this section.

First, the intricate legal systems within the healthcare sector pose a well-acknowledged issue. Additional intricacies in the management of patient data on the blockchain is introduced by the incorporation of the General Data Protection Regulation (GDPR) and other data protection standards. The transparency inherent in blockchain technology may potentially provide challenges in meeting the privacy standards associated with patient data, hence calling for the development of novel solutions that prioritize privacy preservation (Regulation (EU), 2016).

Second, the global regulatory landscape surrounding blockchain tokens exhibits significant variation. The problem lies in effectively navigating the dynamic landscape of token legislation, given that tokens play a pivotal role in numerous crowdfunding models based on blockchain technology (US Security and Exchange Mission, 2023).

Third, network congestion is one important factor to consider in the context of scalability. Blockchain networks encounter challenges in accommodating higher transaction volumes which makes the emergence of scalability difficulties apparent, thus affecting both transaction speed and cost. According to a report from (CoinDesk, n.d.), there is a significant issue of congestion within the Bitcoin blockchain.

The scalability of blockchain networks, particularly those utilizing Proof-of-Work mechanisms, may be impeded by their resource-intensive nature. The problem at hand becomes apparent when considering the computing and energy requirements associated with the process of mining (BitcoinEnergy Consumption Index-Digiconomist, n.d.). For achieving scalability, the establishment of smooth data interchange and system functionality is crucial (Khezr et al., 2019).

3.6 CONCLUSION AND OUTLOOK

The fusion of blockchain technology and IoT has revolutionized healthcare crowdfunding, enhancing transparency, trust, and accessibility. Blockchain technology provides an immutable ledger, allowing donors to see the results of their contributions in real time, boosting trust. Smart contracts automate fund distribution and lower administrative costs, while advanced cryptographic methods protect sensitive patient data. This approach encourages financial inclusion and reduces transaction costs, ensuring more contributions reach the intended beneficiaries. However, regulation-related issues remain crucial to ensure ethical and legal use. The fusion of blockchain and IoT represents a commitment to a society where access to high-quality healthcare is a fundamental right. The future of healthcare crowdfunding is bright, bringing us closer to a more accessible and equitable healthcare system.

REFERENCES

Al-Abbasi, M. F., Baig, Z. A., Anwar, S. M., & Choi, S. (2020). Internet of things (IoT) in health care: A comprehensive survey. *Journal of King Saud University—Computer and Information Sciences*, 20, 435–455.

Bali, V., Bali, S., Gaur, D., Rani, S., & Kumar, R. (2023). Commercial-off-the shelf vendor selection: A multi-criteria decision-making approach using intuitionistic fuzzy sets and TOPSIS. *Operational Research in Engineering Sciences: Theory and Applications*, 6(2).

Bassani, G., Marinelli, N., & Vismara, S. (2019). Crowdfunding in healthcare. *The Journal of Technology Transfer*, 44, 1290–1310. https://doi.org/10.1007/s10961-018-9663-7

Bhambri, P., & Rani, S. (2024). Challenges, opportunities, and the future of industrial engineering with IoT and AI. In *Integration of AI-Based Manufacturing and Industrial Engineering Systems with the Internet of Things* (pp. 1–18). USA: CRC Press.

Bhambri, P., Rani, S., Balas, V. E., & Elngar, A. A. (Eds.). (2023). Integration of AI-Based Manufacturing and Industrial Engineering Systems with the Internet of Things. CRC Press.

BitcoinEnergy Consumption Index-Digiconomist. (n.d.). https://digiconomist.net/bitcoin-energy-consumption

Büyüksaracoğlu, R. T., Elmas, D. B., & Kaya, İ. (2020). A blockchain-based tokenization model for healthcare real estate investments. *Journal of Ambient Intelligence and Humanized Computing*, 11(9), 3781–3790.

Chauhan, M., & Rani, S. (2021). Covid-19: A revolution in the field of education in India. In *Learning How to Learn Using Multimedia* (pp. 23–42). Springer.

Coindesk. (n.d). Bitcoin, Etherium, Crpto news and price data. https://www.coindesk.com

Dal Mas, F., Massaro, M., Verde, J. M., & Cobianchi, L. (2020). Can the blockchain lead to new sustainable business models? *Journal of Business Models*, 8(2), 31–38.

Dexcom. (n.d.). Dexcom continuous glucose monitoring | Dexcom CGM. https://www.dexcom.com/

Fitbit. (n.d.). Fitbit: Find your fit. https://www.fitbit.com/

Ghazal, L. V., Watson, S. E., Gentry, B., & Santacroce, S. J. (2023). "Both a life saver and totally shameful": Young adult cancer survivors' perceptions of medical crowdfunding. *Journal of Cancer Survivorship*, 17(2), 332–341.

Hou, X., Wu, T., Chen, Z., & Zhou, L. (2022). Success factors of medical crowdfunding campaigns: Systematic review. *Journal of Medical Internet Research*, 24(*3*), e30189.

Howe, J. (2006). The rise of crowdsourcing. *Wired*, 14(*6*), 1–4.

Iqbal, M., Yaqoob, I., Anwar, Z., Kazmi, S. A., Imran, M., & Vasilakos, A. V. (2021). Health blockchain: A novel approach for a sustainable environment using IOT and blockchain. *Sustainable Cities and Society*, 75, 103319.

Kataria, A., Agrawal, D., Rani, S., Karar, V., & Chauhan, M. (2022). Prediction of blood screening parameters for preliminary analysis using neural networks. In *Predictive Modeling in Biomedical Data Mining and Analysis* (pp. 157–169). Academic Press.

Kataria, A., Puri, V., Pareek, P. K., & Rani, S. (2023, July). Human activity classification using G-XGB. In *2023 International Conference on Data Science and Network Security (ICDSNS)* (pp. 1–5). IEEE.

Khezr, S., Moniruzzaman, M., Yassine, A., & Benlamri, R. (2019). Blockchain technology in healthcare: A comprehensive review and directions for future research. *Applied Sciences*, 9(*9*), 1736. MDPI AG. http://dx.doi.org/10.3390/app9091736

Kinsa. (n.d.). Kinsa Smart Thermometer. https://www.kinsahealth.co/

Kraus, S., Schiavone, F., Pluzhnikova, A., & Invernizzi, A. C. (2021). Digital transformation in healthcare: Analyzing the current state-of-research. *Journal of Business Research*, 123, 557–567.

Kshetri, N. (2017). Can blockchain strengthen the internet of things. *IT Professional*, 19(*4*), 68–72.

Kumar, P., Banerjee, K., Singhal, N., Kumar, A., Rani, S., Kumar, R., & Lavinia, C. A. (2022). Verifiable, secure mobile agent migration in healthcare systems using a polynomial-based threshold secret sharing scheme with a blowfish algorithm. *Sensors*, 22(*22*), 8620.

Kuo, T. T., Kim, H. E., & Ohno-Machado, L. (2017). Blockchain distributed ledger technologies for biomedical and health care applications. *Journal of the American Medical Informatics Association*, 24(*6*), 1211–1220.

Li, X., Zhang, P., & Zhao, S. (2017). A blockchain based new secure multi-layer network model for IoT. *IEEE Internet of Things Journal*, 4(*4*), 817–828.

Mougayar, W. (2016). *The Business Blockchain: Promise, Practice, and Application of the Next Internet Technology*. Wiley.

Nakamoto, S. (2008). Bitcoin: A peer-to-peer electronic cash system. https://bitcoin.org/bitcoin.pdf

Prabhakar, A., Srivastava, R. M., & Lavanya, S. (2017). Healthcare internet of things (HIoT): A review of medical devices with Bluetooth & BLE. *Procedia Computer Science*, 115, 366–373.

Propeller Health. (n.d.). Propeller health: Home. https://www.propellerhealth.com/

Puri, V., Kataria, A., Solanki, V. K., & Rani, S. (2022, December). AI-based botnet attack classification and detection in IoT devices. In *2022 IEEE International Conference on Machine Learning and Applied Network Technologies (ICMLANT)* (pp. 1–5). IEEE, Salvador.

Rani, S., Bhambri, P., & Kataria, A. (2023a). Integration of IoT, big data, and cloud computing technologies. In *Big Data, Cloud Computing and IoT: Tools and Applications*. CRC Press.

Rani, S., Kataria, A., & Chauhan, M. (2022). Cyber security techniques, architectures, and design. In *Holistic Approach to Quantum Cryptography in Cyber Security* (pp. 41–66). CRC Press.

Rani, S., Kaur, J., & Bhambri, P. (2023b). Technology and gender violence: Victimization model, consequences and measures. In *Communication Technology and Gender Violence* (pp. 1–19). Springer International Publishing.

Rani, S., Kumar, S., Kataria, A., & Min, H. (2023c). SmartHealth: An intelligent framework to secure IoMT service applications using machine learning. *ICT Express*, 10(2), 420–425.

Rani, S., Mishra, A. K., Kataria, A., Mallik, S., & Qin, H. (2023d). Machine learning-based optimal crop selection system in smart agriculture. *Scientific Reports*, 13(*1*), 15997.

Rani, S., Pareek, P. K., Kaur, J., Chauhan, M., & Bhambri, P. (2023e, February). Quantum machine learning in healthcare: Developments and challenges. In *2023 IEEE International Conference on Integrated Circuits and Communication Systems (ICICACS)* (pp. 1–7). IEEE.

Regulation (EU). (2016). European Parliament and of the Council of 27 April 2016 on the protection of natural persons with regard to the processing of personal data and on the free movement of such data, and repealing Directive 95/46/EC GDPR. *Official Journal, L* 119, 1–88. ELI. http://data.europa.eu/eli/reg/2016/679/oj[legislation

Ren, J., Raghupathi, V., & Raghupathi, W. (2020). Understanding the dimensions of medical crowdfunding: A visual analytics approach. *Journal of Medical Internet Research*, 22(*7*), e18813.

Sear, M. C. (2021). *Ethical Issues in Use of Crowdfunding to Finance Healthcare* (Doctoral dissertation, University of Pittsburgh).

Singh, H., & Rani, S. (2023). Flex sensor integrated smart strap to verify correct wearing of the face mask. *IEEE Sensors Journal*, 24(*2*), 2020–2027.

Sivarajah, U., Kamal, M. M., Irani, Z., & Weerakkody, V. (2017). Critical analysis of big data challenges and analytical methods in assessing the quality of healthcare data. In *International Conference on Smart Health* (pp. 216–231). Springer.

Sudevan, S., Barwani, B., Al Maani, E., Rani, S., & Sivaraman, A. K. (2021). Impact of blended learning during Covid-19 in Sultanate of Oman. *Annals of the Romanian Society for Cell Biology*, 14978–14987.

Tama, B. A., Demeester, P., & Moerman, I. (2017). A survey on smart city IoT applications. *Journal of King Saud University-Computer and Information Sciences*, 21, 476–589.

Tanwar, R., Chhabra, Y., Rattan, P., & Rani, S. (2022, September). Blockchain in IoT networks for precision agriculture. In *International Conference on Innovative Computing and Communications: Proceedings of ICICC 2022, Volume 2* (pp. 137–147). Springer Nature Singapore.

US Security and Exchange Mission. (2023). Framework for "investment contract" analysis of digital assets. https://www.sec.gov/corpfin/framework-investment-contract-analysis-digital-assets

WHO. (2021). Public engagement and crowdfunding in health research: A practical guide. Licence: CC BY-NC-SA 3.0 IGO.

Zanella, A., Bui, N., Castellani, A., Vangelista, L., & Zorzi, M. (2014). Internet of things for smart cities. *IEEE Internet of Things Journal*, 1(*1*), 22–32.

Zhang, P., White, J., Schmidt, D. C., Lenz, G., & Rosenbloom, S. T. (2016). FHIRChain: Applying blockchain to securely and scalably share clinical data. *Computers in Biology and Medicine*, 79, 55–64.

Zhang, R., Xue, R., & Liu, L. (2018). Security and privacy in cloud-assisted healthcare systems: A case study of e-health clouds. *IEEE Transactions on Information Forensics and Security*, 13(*6*), 1474–1486.

4 Internet of Things Security

Techniques and Challenges

K. Aditya Shastry and Mohan S.G.

4.1 INTRODUCTION

The advent of the "Internet of Things" (IoT) has brought about a transformative wave of technological advancements, enabling the interconnection of billions of devices worldwide. IoT refers to a sensor-equipped, software-enabled network of real-world items, and connectivity capabilities that permit them to gather and exchange data. This network extends beyond traditional computing devices, encompassing an exhaustive set of objects such as appliances, vehicles, industrial machinery, wearables, and even city infrastructure (Kumar et al., 2019). The rapid proliferation of IoT devices among several areas, involving medical, transportation, manufacturing, and smart homes, has revolutionized the way we live and work. It has facilitated advancements such as remote patient monitoring in healthcare, predictive maintenance in manufacturing, smart energy management in homes, and efficient traffic management in cities. However, the rapid development and widespread adoption of IoT have also brought a host of security challenges and vulnerabilities (Elgazzar et al., 2022).

IoT security is a critical concern due to the interconnected and pervasive characteristics of IoT mechanisms. The distinctive traits of IoT systems make them prone to various security dangers and incidents. The vast number and diversity of IoT tools introduce challenges in implementing robust security measures uniformly across the entire ecosystem (Schiller et al., 2022). Furthermore, numerous IoT devices possess inadequate computational power, memory, and energy constraints, making it difficult to implement resource-intensive security mechanisms (Williams et al., 2022). The security vulnerabilities in IoT systems can have severe consequences. Unlawful access to sensitive information, privacy breaches, device manipulation, and disruption of critical services are among the probable hazards coupled with compromised IoT devices. The inferences of such security breaches extend beyond individual devices to affect entire networks, leading to cascading effects and potential compromises of other connected systems (Kaur et al., 2023).

The need to address IoT security challenges has become increasingly urgent. Governments, regulatory bodies, and industry organizations have recognized the significance of implementing robust security measures to safeguard IoT deployments (Kumar et al., 2022). Efforts are being made to establish standards, guidelines, and best practices to uphold the safety and integrity of IoT tools. Furthermore,

DOI: 10.1201/9781003460367-4

researchers, practitioners, and policymakers are exploring innovative techniques and solutions to mitigate the evolving threats faced by IoT designs and links (Azrour et al., 2021).

Keeping these points in mind, this chapter offers a comprehensive evaluation of the defense landscape in IoT. The chapter aims to thoroughly analyze the security landscape in the environment of IoT systems. It explores the vulnerabilities that have emerged because of the rapid proliferation of IoT mechanisms and the potential threats that compromise the integrity, confidentiality, and availability of IoT approaches. It presents a comprehensive overview of security techniques for IoT. The chapter seeks to deliver a detailed summary of the techniques employed to safe IoT tools and networks. It covers a wide range of security mechanisms, including validation and access control mechanisms, encryption algorithms, secure communication protocols, anomaly detection, and intrusion prevention systems tailored specifically for IoT deployments. It addresses the challenges in implementing robust security measures. The chapter focuses on the challenges faced in implementing strong security measures within IoT systems. It discusses the unique characteristics of IoT devices, such as limited computational power, memory, and energy constraints, which pose significant hurdles to implementing effective security measures. It emphasizes the necessity for a holistic approach to IoT security. The chapter highlights the interconnected nature of IoT gadgets and the importance of adopting a holistic approach to safety. It stresses the importance of considering the complete IoT network, including devices, networks, and backend systems, when designing and employing safety procedures to ensure comprehensive protection. It examines the integration of artificial intelligence (AI) and machine learning (ML) techniques for enhanced risk finding and predictive analysis. Additionally, it explores the influence of blockchain technology in ensuring data integrity and accountability within IoT systems.

Figure 4.1 shows a generic framework for implementing IoT security.

The key components of the IoT security framework are as follows:

- *Device Authentication and Authorization:* This component focuses on ensuring that only legitimate and authorized devices can access the IoT network or communicate with other devices. It involves employing dedicated verification tools, like the unique cryptographic keys or digital certificates, to verify the identity of each device. Additionally, role-based access control can be applied to limit the actions a device can perform based on its specific permissions.
- *Data Encryption:* Information communicated among IoT mechanisms and the central infrastructure should be encrypted to protect it from interception and unauthorized access. Encryption algorithms such as "AES (Advanced Encryption Standard)" are frequently employed to ensure information at rest and in transit.
- *Secure Boot and Firmware Updates:* Securing the device's boot process ensures that only authorized and tamper-free firmware and software can be executed on the IoT tool. Standard firmware updates with security patches are vital to resolve known liabilities and keep devices protected from emerging threats (Bali et al., 2023).

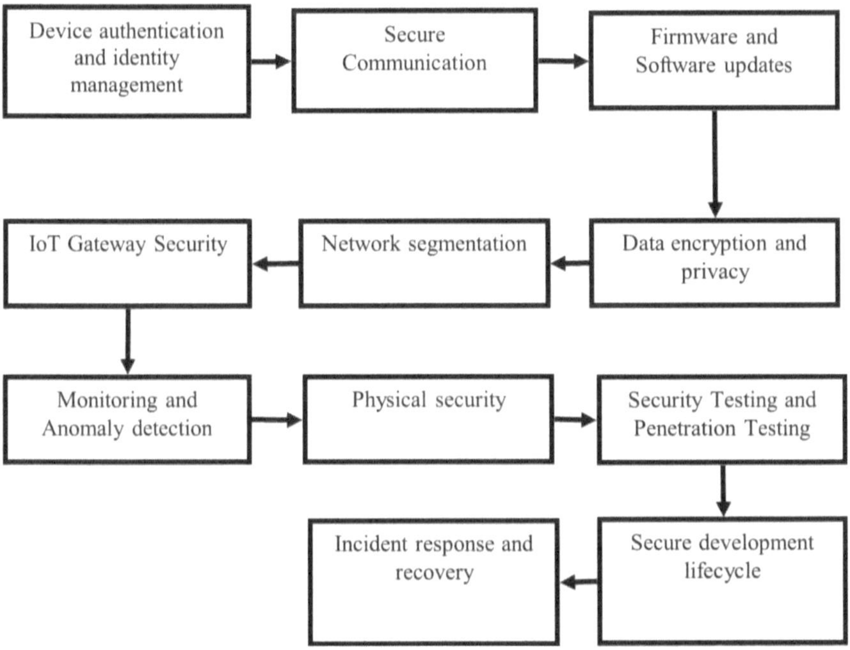

FIGURE 4.1 Generic framework for implementing IoT security.

- *Secure Communication:* The framework should incorporate secure communication protocols such as TLS (Transport Layer Security) or MQTT (Message Queuing Telemetry Transport) with appropriate safety designs. Secure communication prevents eavesdropping, man-in-the-middle attacks, and information tampering during transit (Rani et al., 2022).
- *Physical Security:* Physical security measures should be implemented to safeguard IoT designs from physical tampering or theft. This includes using tamper-resistant hardware, securing device access, and employing anti-tamper mechanisms.
- *Network Security:* Network protection aims at safeguarding the IoT infrastructure, gateways, and communication channels. "Firewalls," "intrusion detection systems (IDS)," and access control methods are instances of grid protection components used to examine and manage network traffic.
- *Device Management and Monitoring:* An effective IoT security framework includes robust device management and monitoring capabilities. This involves tracking devices, managing their life cycle, and monitoring their behavior for any anomalies that could indicate a security breach (Rani et al., 2023c).
- *Privacy Protection:* IoT devices often collect sensitive data, and privacy protection is necessary to guarantee that this data is handled in compliance with appropriate policies as well as user expectations. Implementing data minimization, anonymization, and informed consent practices can help protect user privacy.

- *Security Updates and Patch Management:* The framework should have a process in place for timely security updates and patch management across all IoT devices. Promptly addressing vulnerabilities with updates is crucial to prevent exploitation by malicious actors.
- *Operator Learning and Awareness:* Users and administrators have a crucial part in IoT security. An effective security framework includes educational programs and resources to induce familiarity concerning impending dangers and competent preparations for secure IoT usage.
- *Event Reply and Recovery:* Despite the best security measures, breaches may still occur. A complete event reaction strategy must be part of the framework, outlining steps to detect, respond, and recover from security incidents effectively.
- *Regulatory Compliance:* The IoT security framework should align with relevant industry standards and comply with applicable regulations to ensure legal and regulatory adherence (Ameyed et al., 2023).

By integrating these components, an IoT security framework can significantly enhance the overall safety strength of IoT tools and networks, making them more resilient against potential threats and attacks. Overall, the chapter focuses on delivering a complete insight of the techniques and challenges associated with IoT security. It serves as a helpful source for scholars, specialists, and legislators seeking to comprehend and address the evolving security landscape in the realm of IoT. By addressing these objectives, the chapter contributes to the advancement of protected and dependable IoT ecosystems, ultimately promoting the widespread adoption of IoT technologies.

4.2 IoT SECURITY ENVIRONMENT

This section presents the protection tests in IoT and describes the dangers and damage vectors.

4.2.1 Safety Challenges in IoT

The widespread acceptance of IoT systems has brought forth a variety of safety issues (Rani et al., 2023b). These challenges arise because of the unique characteristics of IoT tools, the complexity of the IoT environment, and the evolving threat landscape. Understanding and addressing these challenges is vital for ensuring the protection and secrecy of IoT deployments. Let's delve into certain key security challenges in IoT (Ali et al., 2022; Tawalbeh et al., 2020; Mazhar et al., 2023):

- *Heterogeneity:* The fundamental challenge in IoT security is the sheer heterogeneity of devices, manufacturers, and communication protocols. IoT encompasses a vast array of tools varying from miniature sensors to large-scale industrial tools, individually with their specific operating system, firmware, and security capabilities. This diversity makes it difficult to establish standardized security practices and creates inconsistencies in implementing security measures across the IoT ecosystem (Chauhan & Rani, 2021).

- *Scalability:* IoT systems are characterized by their massive scale, with billions of connected tools worldwide. This scale presents unique security challenges as securing each device and managing security at such magnitude becomes a daunting task. Additionally, as the tools increase, so does the attack surface, providing more opportunities for cybercriminals to exploit vulnerabilities and gain unauthorized entry to the IoT network.
- *Limited Resources:* Several IoT tools exhibit restricted computational power, memory, and energy resources. Such source restrictions frequently lead to the development of lightweight operating systems and firmware, which may lack robust security features. Furthermore, the limited resources make it challenging to apply regular security updates and patches, leaving devices vulnerable to known exploits and vulnerabilities.
- *Legacy Systems:* Incorporating IoT tools with the prevailing legacy systems and infrastructure introduces security challenges. Legacy systems may have outdated or unsupported security protocols, making them susceptible to attacks. Additionally, compatibility issues between legacy systems and newer security practices can create weak links in the IoT ecosystem, allowing attackers to exploit vulnerabilities in older components.
- *Inadequate Authentication and Authorization:* Weak or inadequate authentication mechanisms pose a significant challenge in IoT protection. Several IoT gadgets rely on default or easily guessable credentials, making them vulnerable to brute-force attacks and unauthorized access. Insufficient authorization controls can allow unauthorized users to gain privileged approach to restricted information or control critical devices, leading to data breaches or malicious manipulation of IoT systems.
- *Data Privacy:* Privacy concerns are paramount in IoT deployments. IoT tools gather enormous volumes of susceptible information, incorporating individual data, health records, and location data. Ensuring the privacy and confidentiality of this information across its development, including collection, transmission, storage, and processing, is crucial. Inadequate encryption, insecure data transmission protocols, and weak access controls may cause illegal access, data breaches, and violations of privacy regulations.
- *Network Security:* Securing the communication networks within the IoT ecosystem is vital. IoT networks typically consist of various interconnected components, including sensors, gateways, cloud platforms, and backend systems. Weak network security measures can expose these communication channels to eavesdropping, interception, and unauthorized access. A compromised network component might operate as a gateway for attackers to infiltrate the complete IoT infrastructure, compromising the reliability and secrecy of data.
- *Supply Chain Vulnerabilities:* The complex supply chain involved in manufacturing and deploying IoT devices introduces potential vulnerabilities. Malicious actors can exploit weaknesses in the supply chain, such as compromised components or preinstalled malware, to acquire unlawful approach to IoT devices or compromise their security. It is vital to implement strong supply chain administration procedures to guarantee the reliability and security of IoT devices from the manufacturing stage to deployment.

- *Evolving Threat Landscape:* The threat landscape in IoT is constantly evolving as attackers adapt and discover new vulnerabilities. IoT tools may become targets for several kinds of attacks, containing malware infections, DDoS attacks, botnet exploitation, and physical tampering. The rapid pace of technological advancements, linked with the lack of standardized security practices, makes it challenging to stay ahead of emerging dangers and liabilities.

Addressing these security challenges requires a comprehensive and multilayered approach. It encompasses executing safety procedures at the device level, securing network infrastructure, establishing standardized security frameworks, promoting secure development practices, and fostering collaboration among stakeholders. By addressing these challenges proactively, the IoT ecosystem can become more resilient and trustworthy, ensuring the continued growth and benefits of IoT technology (Puri et al., 2022).

4.2.2 Dangers and Damage Vectors

The interconnected nature of IoT tools along with the huge volumes of sensitive records they generate make them desirable objects for malicious actors. Understanding the threats and attack vectors in the IoT landscape is crucial for implementing effective security measures. Here are certain general threats and attack vectors that pose risks to IoT systems (Khan et al., 2022; Shafiq et al., 2022; Affia et al., 2023):

- *Unauthorized Access:* Unauthorized access is a significant threat in IoT deployments. Attackers may exploit weak or default credentials, improper authentication mechanisms, or unpatched weaknesses to obtain unlawful approach to IoT devices or networks. Once inside, they can manipulate device settings, steal sensitive data, or launch further attacks within the IoT ecosystem.
- *Malware and Ransomware:* IoT devices can be vulnerable to malware infections, where malicious software is installed without the device owner's knowledge. Infected devices can become part of a botnet, such as the Mirai botnet, that could be utilized to launch large-scale distributed denial-of-service (DDoS) attacks, disrupting network services. Ransomware attacks targeting IoT devices are also on the rise, where attackers encrypt device data and demand a ransom for its release.
- *Data Interception and Manipulation:* IoT devices collect and transmit susceptible information, incorporating private data, health records, and industrial data. Inadequate encryption or weak data transmission protocols can expose this data to interception, eavesdropping, and tampering. Attackers may intercept data to gain illegal access to protected information, manipulate data to deceive systems or users, or launch attacks by injecting malicious commands into data streams.
- *Physical Attacks:* Physical attacks on IoT devices pose a significant threat. Hackers may tangibly manipulate the devices, such as removing

or modifying sensors, altering device functionality, or manipulating data. Physical attacks may cause false alarms, disruptions in critical processes, or unauthorized access to systems and data. Additionally, attacks on the physical infrastructure supporting IoT systems, such as power supply or network infrastructure, can result in service interruptions and compromise the overall security of IoT deployments.

- *Supply Chain Vulnerabilities:* The complex supply chain involved in manufacturing and deploying IoT devices introduces potential vulnerabilities. Attackers may exploit weaknesses in the supply chain to compromise the security of devices before they even reach the end-users. This incorporates the insertion of malicious components, compromised firmware, or unauthorized modifications to the devices. Supply chain incidents can compromise the integrity and protection of IoT devices, permitting hackers to obtain unlawful admission or regulate the devices or the entire IoT infrastructure.
- *"Denial-of-Service (DoS)" and "Distributed Denial-of-Service (DDoS)" Threats:* These attacks can disrupt IoT services by overwhelming the network or devices with an excessive amount of traffic, rendering them unable to function properly. IoT devices, with their limited resources, can be easily overwhelmed by such attacks. As a result, critical services may be unavailable, causing financial loss, safety risks, or interruptions in essential operations (Rani et al., 2023d).
- *"Man-in-the-Middle (MitM)" Attacks:* MitM attacks occur when attackers capture and control interaction across IoT devices or between devices and backend systems. By eavesdropping on the transmission passage, hackers may obtain access to restricted data, inject malicious commands, or manipulate data, leading to illegal actions or compromised system integrity. Weak encryption, vulnerable rules, or hacked system elements can facilitate MitM attacks.
- *Manipulation of Social Contexts:* Social engineering techniques, like phishing, are used by the hackers to cheat the consumers into uncovering private facts or yielding unlawful approach to IoT devices. Attackers may send fraudulent emails, messages, or make phone calls posing as legitimate entities to deceive users and gain their trust. Once successful, hackers are able to manipulate the obtained information to compromise the safety of IoT systems.
- *Firmware and Software Exploits:* Vulnerabilities in firmware or software used in IoT tools could be manipulated by hackers to secure illegal access, manipulate device behavior, or inject malicious code. Unpatched vulnerabilities or outdated firmware/software versions provide opportunities for attackers to exploit weaknesses and compromise the security of IoT tools and the entire network.

Addressing these dangers and damage vectors requires a multifaceted approach. This includes implementing robust authentication and authorization mechanisms, encrypting data, regularly patching and updating firmware/software, monitoring network traffic for anomalies, educating users about recommended safety precautions, and encouraging a culture of security and vigilance among all stakeholders inside the IoT ecosystem. By proactively addressing these threats, the safety and reliability of IoT deployments can be significantly enhanced (Rani et al., 2023a).

4.3 VERIFICATION AND EDIT CONTROL

Verification and edit regulation procedures have a vital part in ensuring the safety and reliability of IoT systems. These mechanisms help verify the identity of users or devices and control their access to IoT resources, thus stopping unlawful access and guarding restricted information (Singh & Singh, 2023).

4.3.1 User and Device Authentication

User authentication is essential in IoT systems in which the individual consumers communicate with IoT devices or access IoT services (Rani et al., 2023a). It guarantees that the authenticated personnel can access and control IoT devices or access sensitive data generated by these devices. Some common user authentication techniques used in IoT are as follows (Uppuluri & Lakshmeeswari, 2023; Taherdoost, 2023; Ragothaman et al., 2023; Samaila et al., 2018):

- *Username and Password:* This forms the most occurring form of user authentication, where users provide a unique username and a corresponding password. It is important to enforce strong password policies, such as requiring an amalgamation of alphanumeric characters, special symbols, and regular password updates. Additionally, the usage of two-factor authentication (2FA) or "multi-factor authentication" (MFA) will boost the safety of user authentication by combining passwords with additional authentication factors like SMS codes, biometrics, or hardware tokens.
- *Biometrics:* Biometric authentication methods, like fingerprint recognition, iris scanning, or facial recognition, could be employed to authenticate the identity of users. Biometric data is unique to each individual, making it a secure form of authentication. However, the implementation of biometric authentication in IoT systems may require additional hardware and processing capabilities on the user's device or within the IoT infrastructure.
- *Certificates:* Digital certificates can be used for user authentication in IoT systems. Certificates bind a user's identity to a public key and are issued by trusted certification authorities (CAs). Users present their certificates during authentication, and the validity and authenticity of the certificate are verified by the IoT system. Certificate-based authentication provides strong security and could be combined with additional verification techniques for enhanced security.

Device authentication is crucial in IoT deployments to guarantee that just the authenticated devices could use IoT assets or communicate with other devices or backend systems. It blocks unlawful tools from obtaining control over IoT infrastructure or tampering with sensitive data. Here are some common device authentication techniques used in IoT (Joshi et al., 2021; Sidhu et al., 2019; Litvinov et al., 2023; Abbas et al., 2021):

- *Pre-Shared Keys (PSK):* PSK authentication involves sharing a secret key between the device and the IoT system. The device presents the key during authentication, and if it matches the expected value, the device is considered authenticated. PSK authentication is simple and suitable for resource-constrained devices, but it requires secure key distribution and management processes.
- *Public Key Infrastructure (PKI):* PKI-based authentication uses public–private key pairs to authenticate devices. Each device possesses a unique private key, while the corresponding public key is employed for verification. The IoT system verifies the device's identity by validating its digital certificate, which is issued by a trusted CA. PKI provides strong authentication and enables secure communication between devices and IoT systems (Sudevan et al., 2021).
- *Secure Element:* Secure elements, such as hardware security modules (HSMs) or trusted platform modules (TPMs), can be used for device authentication. These dedicated hardware components store cryptographic keys and perform authentication functions securely. Secure elements provide tamper-resistant storage for device credentials and ensure the integrity of authentication processes.
- *Mutual Authentication:* Mutual authentication involves the authentication of both the device and the IoT system. This ensures that both entities verify each other's identity before establishing a trusted communication channel. Mutual authentication provides a superior degree of security, protecting against man-in-the-middle attacks and unauthorized device impersonation.

Table 4.1 provides a comparison of the user authentication and device authentication procedures with respect to various criteria.

Certain best practices which could be followed for User and Device Authentication are highlighted below:

- *Strong Credential Management:* Impose robust password strategies for user authentication, and securely store device credentials such as keys, certificates, or shared secrets.
- *Regular Updates:* Keep user authentication mechanisms and device firmware/software equipped with the modern protection reinforcements and updates to address any known vulnerabilities.
- *Multi-Factor Authentication:* Implement multi-factor authentication for consumers to add an extra layer of security beyond passwords.
- *Secure Key Distribution:* Establish secure mechanisms for distributing and managing device credentials, such as using secure provisioning processes during device onboarding.
- *Certificate Life Cycle Management:* Implement proper certificate life cycle management processes, including certificate issuance, revocation,

TABLE 4.1
Comparison of User and Device Authentication Methods

Authentication Method	User Authentication	Device Authentication
Purpose	Verify the identity of human users	Verify the identity of IoT devices
Authentication factors	Username, password, biometrics, certificates, etc.	Pre-shared keys, public–private key pairs, certificates, etc.
Security considerations	Password policies, multi-factor authentication (MFA), secure storage of credentials, secure communication	Secure key distribution, certificate life cycle management, secure storage (HSMs, TPMs), secure communication
Implementation complexity	Relatively straightforward, involving user input and verification	May require additional hardware or secure elements for key storage and authentication processes
Associated risks	Password breaches, unauthorized access by users, social engineering attacks	Device impersonation, unauthorized device access, tampering or manipulation of device data
Example techniques	Username and password, biometrics, certificates, two-factor authentication (2FA)	Pre-shared keys (PSK), public key infrastructure (PKI), secure elements (HSMs, TPMs)

and renewal, to ensure the integrity and validity of certificates used in authentication.

- *Secure Communication:* Utilize protected transmission procedures like "TLS (Transport Layer Security)" or SSH (Secure Shell) to protect authentication processes and data transmission.
- *Monitoring and Auditing:* Implement monitoring and auditing mechanisms to identify and react to suspicious authentication attempts or unauthorized access activities.

By employing strong user and device authentication mechanisms and following best practices, IoT systems can enhance security, prevent unauthorized access, and protect sensitive data generated by IoT devices.

4.3.2 Role of Public Key Infrastructure

"Public Key Infrastructure" (PKI) plays a fundamental role in ensuring the safety and credibility of the IoT ecosystem. With the ever-increasing number of connected devices and the criticality of data exchanged within IoT networks, PKI provides a robust framework to establish secure communication, data integrity, and authentication mechanisms (Mathur & Arora, 2020; Ullah et al., 2021; Javaid et al., 2020; Khalil et al., 2022; Liu et al., 2021):

- *Device Authentication and Identity Management:* In IoT, billions of devices, including sensors, actuators, and other smart objects, communicate and exchange data autonomously. Ensuring the authenticity of these devices is crucial to prevent unauthorized access, data tampering, and potential security breaches. PKI enables device authentication by issuing unique digital certificates to each IoT device. These certificates contain the device's public key and other identity information, signed by a trusted Certificate Authority (CA). When devices communicate, they use their private keys to sign messages, allowing other devices or servers to verify their authenticity using the corresponding public keys. This process establishes a chain of trust, ensuring only authorized and verified devices can participate in the IoT network.
- *Data Encryption and Confidentiality:* PKI facilitates protected interaction among IoT tools and gateways via the usage of encryption. When devices exchange sensitive data, PKI-based encryption ensures that only the intended recipient can decipher the information using its private key. This prevents unauthorized entities from intercepting and reading sensitive data, maintaining data confidentiality within the IoT ecosystem.
- *Data Integrity and Non-Repudiation:* Data integrity is vital in IoT, as any tampering or alteration of data can lead to inaccurate decisions and potential harm. PKI supports digital signatures, which use the sender's private key to create a unique signature for each message. Upon receipt, the addressee can validate the signature using the sender's public key, confirming that the data was not manipulated with during transmission. Additionally, digital signatures provide non-repudiation, meaning that the transmitter cannot decline having shown a specific message, as their private key is utilized to sign it.
- *Secure Over-the-Air (OTA) Updates:* IoT devices often require software updates to patch vulnerabilities and add new features. PKI is instrumental in securing OTA updates. When a device receives an update, the authenticity and integrity of the update package could be validated using digital signatures. PKI guarantees that the authenticated and verified updates are accepted, lowering the danger of malicious code injection and improving the general safety of the IoT network.
- *Secure Device Enrollment:* The procedure of enrolling new devices into an IoT network must be secure to prevent the inclusion of unauthorized or compromised devices. PKI enables secure device enrollment by ensuring that new devices are issued valid digital certificates only after proper authentication and authorization. This process guarantees that devices joining the IoT network are legitimate and can be trusted.
- *Revocation and Key Management:* In the dynamic IoT landscape, devices may be lost, stolen, or compromised. PKI includes mechanisms for certificate revocation and key management. If a device's private key is compromised or if a device ought to be decommissioned, its digital certificate could be retracted by the CA, rendering it invalid. This ensures that compromised or unauthorized devices cannot participate in the IoT network even if they possess a valid certificate.

4.3.3 Access Management Systems

These mechanisms are fundamental components of data safety systems that regulate and manage consumer access to resources, data, and services. These mechanisms demonstrate a crucial part in safeguarding restricted data, avoiding illegal access, and ensuring compliance with security policies. There are numerous access regulation mechanisms, each tailored to specific security requirements and scenarios. Let us delve into some of the key access control mechanisms (Duan et al., 2022; Dolgov et al., 2022; Yu et al., 2023; Hou et al., 2020; BenMbarak et al., 2023; Alramadhan & Sha, 2017; Sankaran et al., 2019):

- *"Discretionary Access Control (DAC)":* DAC allows data owners or administrators to permit or limit entry rights to resources centered on the discretion of the owner. In DAC systems, each resource (e.g., files, folders, documents) has an associated Access Control List (ACL), which contains a listing of users or groups and their corresponding access permissions (e.g., read, write, execute). The authorized staff have full power over their resources and can grant or revoke permissions to others. While DAC offers flexibility, it can lead to security risks if owners are not vigilant in managing access rights, potentially allowing unauthorized consumers to access sensitive data.
- *"Mandatory Access Control" (MAC):* MAC enforces a centralized security policy that governs access to resources based on predefined rules and classifications. Unlike DAC, where owners determine access, MAC is controlled by a security administrator or system policy. MAC systems assign security labels (e.g., sensitivity levels) to resources and users. Access choices are done by comparing the safety marker of the source and the client attempting to access it. This ensures a higher level of control and prevents users from overriding access rights, making MAC suitable for environments with strict security requirements.
- *Role-Based Access Control (RBAC):* RBAC organizes users into predefined roles, and access privileges are united with these roles rather than individual users. Operators are assigned to specific parts centered on their job responsibilities, and their entry authorizations are controlled by the roles they are assigned. RBAC simplifies access management by reducing the volume of access control entries and streamlining user provisioning. It also enhances security by ensuring that operators just retain access to resources necessary for their roles, minimizing the danger of excessive privileges.
- *Attribute-Based Access Control (ABAC):* ABAC evaluates access decisions based on multiple attributes related to users, resources, and the environment. These attributes can include user roles, location, time of day, device type, and other contextual information. ABAC allows for more fine-grained and dynamic access control, enabling organizations to enforce complex access policies tailored to specific situations. This process is specifically effective in dynamic and heterogeneous environments like the IoT.
- *Rule-Based Access Control:* It uses a series of predefined policies to verify read permissions. These rules are usually built on specific conditions

or attributes, and access is allowed or blocked based on whether these conditions are met. It might be utilized in combination with additional access regulation methods to enforce additional constraints and conditions.
- *Attribute-Based Encryption (ABE):* ABE is a cryptographic access control mechanism that encrypts information in a manner that only users possessing specific attributes can decrypt and access it. It aligns with the ABAC model and allows for more flexible and dynamic access control, especially in cloud computing and data-sharing scenarios (Tanwar et al., 2022).

4.4 ENCRYPTION TECHNIQUES FOR IoT

Encryption techniques are essential for ensuring the privacy, reliability, and authenticity of data transmitted and stored in IoT devices and networks. As IoT tools regularly possess inadequate sources and are deployed in diverse environments, selecting appropriate encryption methods is vital to balance security and efficiency. In this segment, we examine three main types of encryption techniques for IoT:

4.4.1 Symmetric and Asymmetric Encryption

Symmetric encryption is vital in securing data and communications within the IoT ecosystem. With the proliferation of interconnected IoT tools and the sensitive nature of data transmitted between them, implementing robust encryption mechanisms is essential to protect against potential security threats and illegal access. The key characteristics of symmetric encryption in IoT are as follows (Tariq et al., 2023; Alagheband & Mashatan, 2022; Mosenia & Jha, 2017; Harbi et al., 2021; Zhang et al., 2020):

- *Efficient Resource Usage:* IoT tools possess constrained computational power, memory, and energy resources. Symmetric encryption algorithms are preferred in IoT environments because of their effectiveness and speed. Such procedures need fewer resources, making them well-suited for resource-constrained devices.
- *Data Confidentiality:* One of the primary goals of encryption in IoT is to ensure data confidentiality. When sensitive data is transmitted between IoT devices or to centralized servers, symmetric encoding ensures that just the permitted parties having the joint undisclosed key can decipher and access the information.
- *Secure Communication:* In IoT networks, the tools connect with each other to exchange information and perform various tasks. By encrypting the data using symmetric encryption, IoT devices can securely interchange information over hypothetically unsafe communication channels, safeguarding against eavesdropping and man-in-the-middle attacks.
- *Protected Firmware Updates:* IoT tools regularly need package updates to manage liabilities and enhance functionality. Symmetric encryption can be employed to securely distribute firmware updates, guaranteeing that just authentic updates are accepted and applied by devices.

- *End-to-End Security:* Symmetric encryption can provide end-to-end security between IoT devices and cloud servers or gateways. Data encrypted at the device level remains encrypted until it reaches its final destination, reducing the threat of exposure to files during transmission and storage (Rani et al., 2023a).
- *Key Management:* Proper key management is critical in symmetric encryption. In an IoT deployment, managing secret keys securely is a challenging task, especially if devices are dispersed geographically or subject to physical access by potential adversaries. Ensuring that keys are securely stored and exchanged is vital to maintaining the whole protection of the IoT system (Kataria et al., 2023).

While symmetric encryption offers efficiency and speed, there are some challenges and considerations when implementing it in IoT environments (Hamza et al., 2019; Yang et al., 2022; Lv, 2022):

- *Key Distribution:* Distributing secret keys securely to each IoT device is critical. Any compromise or leakage of the secret key could lead to unauthorized access and data breaches. Secure key distribution mechanisms, such as the Diffie–Hellman key exchange or utilizing a reliable third party for managing critical resources, must be employed.
- *Key Rotation:* Periodic key rotation is essential to enhance security and reduce the adverse influence of a potential key compromise. In IoT, managing key rotation can be more challenging because of the large number of gadgets and potential communication disruptions during key updates.
- *Limited Processing Power:* Some symmetric encryption algorithms might still be computationally expensive for certain low-power IoT devices. Using lightweight cryptographic algorithms explicitly devised for IoT products can help address this concern.
- *Secure Storage:* Storing the secret key securely within each IoT device is critical to prevent physical attacks or unlawful approach to the retrieve the key.

Asymmetric encryption, also known as public key cryptography, plays a critical role in securing data and communications within the IoT ecosystem. It addresses certain issues associated with symmetric encryption in IoT, such as key distribution and safe transfer across devices. Asymmetric encryption uses a pair of keys—public and private keys—that are accurately linked but computationally infeasible to derive one from the other.

The key components of asymmetric encryption in IoT (Henriques & Vernekar, 2017; López Delgado et al., 2022) are as follows:

- *Public Key:* The public key is widely distributed and known to everyone in the IoT network. It is employed for encoding files which could only be deciphered using the related private key. Public keys are utilized to perform secure communication channels, verify digital signatures, and encrypt files transmitted from the IoT tools to other devices or servers.

- *Private Key:* The private key is kept confidential and is known only to the owner of the public–private key pair. It is employed for decrypting data coded with the related public key and generating digital signatures. Private keys must be securely stored to avoid unlawful access, as the compromise of a private key could lead to a breach of security.

Key characteristics of asymmetric encryption in IoT (Silva et al., 2023) are as follows:

- *Key Distribution:* Asymmetric encryption reduces the necessity for a secure key exchange process that forms a significant challenge in symmetric encryption. The public keys can be freely distributed among IoT devices, while the private keys are kept secure and never shared.
- *Secure Communication:* Asymmetric encryption enables secure interaction across the IoT tools minus the necessity of a pre-shared secret key. Devices can use each other's public keys to encrypt messages and send them over potentially unsecured communication channels, ensuring that just the expected recipients having the related private keys can decode and access the information.
- *Digital Signatures:* In IoT applications, digital signatures are crucial for data integrity and authentication. Devices may employ their private keys to generate digital signatures for the files they transmit. Recipients can then verify the legitimacy and reliability of the data using the sender's public key (Rani et al., 2023a).
- *Secure Device Enrollment:* Asymmetric encryption may be employed during the device enrollment process in IoT networks. Devices may employ their private keys to sign registration requests, providing proof of their authenticity to the central server or gateway. This helps prevent unauthorized or rogue devices from joining the IoT network.
- *Secure Firmware Updates:* Asymmetric encryption can also be utilized to securely distribute firmware updates to IoT devices. The updates can be digitally signed by the manufacturer using their private key, and devices can verify the authenticity of the updates using the manufacturer's public key.

Regardless of its advantages, certain challenges and considerations when implementing asymmetric encryption in IoT are highlighted below:

- *Computational Overhead:* Asymmetric encryption is computationally more intensive than symmetric encryption, which might be a concern for resource-constrained IoT tools with reduced administering capacity and energy resources.
- *Key Management:* While asymmetric encryption eliminates the challenge of key distribution, managing a large number of public–private key pairs in a distributed IoT network can be complex and require efficient key management systems (Bhambri & Rani, 2024).
- *Hybrid Approaches:* In practice, hybrid encryption approaches are regularly employed in IoT, combining the efficiency of symmetric encryption for

information broadcast with the security benefits of asymmetric encryption for protected key exchange and authentication.

4.4.2 Lightweight Encryption Algorithms

Lightweight encryption algorithms are specifically designed to tackle the resource constraints of IoT devices. As IoT tools possess constrained processing power, memory, and energy resources, traditional encryption algorithms may be impractical because of their computational complexity and memory requirements. Lightweight encryption algorithms are optimized to maintain equilibrium among security and efficiency, making them well-suited for securing data and communications in IoT environments. Certain key features and examples of lightweight encryption algorithms used in IoT are summarized below (Meng & Buchanan, 2020; Hassan, 2021):

- *Low Computational Overhead:* Lightweight encryption algorithms are developed to minimize computational overhead, enabling efficient encryption and decryption on resource-constrained IoT devices. They utilize straightforward numerical procedures and smaller key sizes to achieve faster processing (Rani et al., 2023e).
- *Low Memory Footprint:* These procedures are devised to use minimal memory, reducing the strain on limited memory resources in IoT tools. This is achieved through compact data structures and efficient memory management techniques.
- *Energy-Efficient:* IoT devices are often powered by batteries, so energy efficiency is crucial to prolong battery life. Lightweight encryption algorithms optimize energy consumption by reducing the number of computations and minimizing the amount of data processing.
- *High Security:* Despite their lightweight nature, these algorithms aim to provide a satisfactory level of security. They employ various cryptographic techniques to ensure data secrecy, reliability, and authenticity (Bhambri et al., 2023).

Examples of Lightweight Encryption Algorithms in IoT:

- *AES-CCM (Advanced Encryption Standard with Counter with CBC-MAC):* AES-CCM is a widespread lightweight encoding procedure that combines AES for encryption with CCM for authentication. Because of its ease of use and effectiveness, it has found widespread use in low-resource IoT gadgets. AES-CCM provides both confidentiality and veracity of information, making it suitable for protected information communication in IoT systems (AlJabri et al., 2023).
- *LEA (Lightweight Encryption Algorithm):* LEA is a symmetric encryption algorithm designed to offer efficient and secure encryption for lightweight applications. It comprises a simple structure with minimal memory needs and is optimized for high-speed encoding and decipherment in controlled IoT tools (Lee et al., 2014).

- *SIMON and SPECK:* SIMON and SPECK are families of lightweight block ciphers developed by the US National Security Agency (NSA). They have small key sizes and compact implementations, making them well-suited for IoT products having reduced assets (Beaulieu et al., 2015).
- *Tiny Encryption Algorithm (TEA):* TEA is a simple and lightweight symmetric encryption procedure which could be implemented in as little as 32 bytes of code. It is used in resource-constrained devices and applications where code size and memory usage are critical factors (Al-Ajarmah, 2023).

4.4.3 Lightweight Encryption Algorithms

Key management in resource-constrained devices is a critical aspect of guaranteeing the safety and confidentiality of information in the IoT ecosystem. As IoT tools possess restricted processing power, memory, and energy resources, implementing efficient and secure key management practices becomes even more challenging. Proper key management is necessary to avoid unlawful access, data breaches, and prospective protection dangers. Certain crucial factors and techniques for key management in resource-constrained IoT tools are summarized below (Taurshia et al., 2022; Gautam & Kumar, 2021):

- *Key Generation and Storage:* Generating strong cryptographic keys is the fundamental stage in key management. Resource-constrained mechanisms may not have access to hardware-built arbitrary number originators, which are best for creating cryptographic keys. In such cases, pseudo-random number generators (PRNGs) may be used, but it is important to ensure their quality and unpredictability. The storage of cryptographic keys is critical. Keys should be stored securely in non-volatile memory to survive device reboots or power cycles. Hardware security modules (HSMs) or secure elements may be employed to provide hardware-based protection for cryptographic keys.
- *Key Distribution:* Distributing secret keys securely to resource-constrained devices can be challenging (Kataria et al., 2022). In cases where devices are physically accessible, attackers may attempt to extract keys from memory or during transmission. The following are some of the techniques for secure key distribution:
  - *Pre-provisioning:* Factory-installed keys can be provisioned during manufacturing and strongly accumulated in the device (Singh & Rani, 2023).
  - *Over-the-Air (OTA) Provisioning:* Using secure communication channels, keys can be provisioned and updated remotely via encrypted OTA updates.
  - *Key Establishment Protocols:* Cryptographic protocols like Diffie–Hellman key exchange can be used to establish shared keys between devices securely.
- *Key Revocation and Rotation:* Key revocation and rotation are essential for maintaining the protection of IoT tools. In the consequence of a key compromise or a device being decommissioned, the associated keys should

be revoked and replaced with new ones. Key rotation is also important to limit the impact of a compromised key. However, rotating keys in resource-constrained devices must be carefully planned to avoid excessive resource consumption and potential service disruptions.

- *Key Management Protocols:* Lightweight key management protocols specifically designed for resource-constrained devices can simplify and optimize key management. Protocols like CoAP (Constrained Application Protocol) with "DTLS (Datagram Transport Layer Security)" can enable secure communication and key management in constrained environments.
- *Secure Boot and Firmware Updates:* Ensuring the reliability of the device firmware is critical for secure key management. Devices should employ secure boot mechanisms to avoid unlawful modifications to firmware. Secure firmware updates may need to be utilized to update cryptographic keys and ensure the latest security measures are in place.
- *Physical Security:* Physical security of resource-constrained devices is crucial. Measures such as tamper-resistant packaging, secure enclosures, and protecting devices from physical attacks can prevent unauthorized access to cryptographic keys.

4.5 THE TRANSFORMATIVE INTERSECTION OF 5G TECHNOLOGY AND IoT

The merging of 5G technology and the IoT has become a focal point of technological innovation in recent years, sparking groundbreaking advancements in connectivity and smart applications. This section analyzes four relevant publications that shed light on the transformative impact of 5G on IoT and the reshaping of our digital future (Rani et al., 2023a).

The paper authored by Pons et al. (2023) explores the utilization of 5G technologies in IoT applications and highlights the challenges caused by interference and network optimization difficulties. The authors emphasize the importance of addressing these issues to ensure reliable and efficient connectivity for IoT tools. The potential gains of the conjunction of networks and services are also examined, having specific focus on 5G's role in enabling a variety of new and innovative applications, including those in smart cities.

Khan et al. (2022) discuss the enhancement of wireless transmission equipment in the framework of 5G networks, which enable the design of IoT ecosystems. The authors outline various applications, such as smart cities, agriculture, retail, and transportation systems, while also addressing challenges related to energy efficiency, security, and performance. The work offers helpful insights for future studies on communication technologies for IoT in the 5G era.

In Hu et al. (2021), a self-powered 5G NB-IoT system for remote monitoring applications, driven by a magnetic-assisted noncontact energy harvester (MN-EH), is presented. This system effectively converts wind power into electrical power, enabling long-range transmission of temperature and CO concentration data to a cloud server. The paper addresses challenges related to power consumption, communication range, and operation lifespan, providing an effective strategy for self-powered remote monitoring systems.

These reviewed publications collectively demonstrate the transformative potential of the convergence of 5G and IoT technologies. They highlight the issues and prospects linked with this synergy, including interference and network optimization difficulties, communication technology development, MTC support in 5G, and self-powered remote monitoring solutions. The research presented in these articles contributes significantly to advancing the seamless integration of 5G and IoT, ushering in a future of enhanced connectivity, intelligence, and innovation across various industries and applications.

4.6 CONCLUSION AND FUTURE DIRECTIONS

This chapter offers a comprehensive analysis of the security landscape in IoT, highlighting both the techniques employed to secure IoT systems and the challenges faced in implementing robust security measures. The rapid proliferation of IoT devices across various sectors has brought unprecedented opportunities and conveniences. However, it has also exposed significant security vulnerabilities, making it crucial to address the integrity, confidentiality, and availability of IoT systems. The chapter begins by discussing the unique characteristics of IoT tools which render them vulnerable to cyber threats, such as diverse device types and their inherent resource limitations. It emphasizes the interconnected nature of IoT devices, necessitating a holistic approach to security that encompasses the entire IoT ecosystem. To tackle these challenges, the chapter explores various security techniques tailored for IoT deployments. It delves into authentication and access control mechanisms, encryption algorithms, and secure communication protocols that safeguard information communication across the IoT tools and backend systems. Additionally, it examines the importance of anomaly detection and intrusion prevention systems in enabling real-time threat detection and mitigation within IoT environments. The paper also highlights the role of policy and regulations in promoting IoT security. It underscores the importance of industry standards, best practices, and collaboration between stakeholders to establish a secure and resilient IoT infrastructure.

The future scope for IoT security is vast and promising, driven by the continuous growth and adoption of IoT devices across various industries. Advancements in artificial intelligence and machine learning will enable advanced threat detection and prevention, while quantum-resistant cryptography will safeguard against emerging quantum-based threats. Enhanced authentication mechanisms, privacy-preserving techniques, and blockchain integration will protect user data and ensure data integrity. Securing edge computing and cloud services in IoT deployments will be crucial, and regulatory frameworks and industry standards will promote a culture of security. Secure device life cycle management and IoT security automation will streamline operations, empowering organizations to build secure and trustworthy IoT systems that prioritize innovation while safeguarding user data and privacy.

REFERENCES

Abbas, S. G., Vaccari, I., Hussain, F., Zahid, S., Fayyaz, U. U., Shah, G. A., Bakhshi, T., & Cambiaso, E. (2021). Identifying and mitigating phishing attack threats in IoT use cases using a threat modelling approach. *Sensors (Basel),* 21(14), 4816. https://doi.org/10.3390/s21144816.

Affia, A.-O., Finch, H., Jung, W., Samori, I. A., Potter, L., & Palmer, X.-L. (2023). IoT health devices: exploring security risks in the connected landscape. *IoT*, 4(2), 150–182. https://doi.org/10.3390/iot4020009.

Alagheband, M. R., & Mashatan, A. (2022). Advanced encryption schemes in multi-tier heterogeneous internet of things: taxonomy, capabilities, and objectives. *Journal of Supercomputing*, 78, 18777–18824. https://doi.org/10.1007/s11227-022-04586-1.

Al-Ajarmah, M. (2023). Exploring the tiny encryption algorithm: a comparative analysis of parallel and sequential computation. *International Journal of Scientific and Engineering Research*, 8, 693. https://doi.org/10.1729/Journal.35208.

Ali, A., Mateen, A., Hanan, A., & Amin, F. (2022). Advanced security framework for IoT. *Technologies*, 10(3), 60. https://doi.org/10.3390/technologies10030060.

AlJabri, Z., Abawajy, J., & Huda, S. (2023). A comprehensive review of lightweight authenticated encryption for IoT devices. *Wireless Communications and Mobile Computing*, 2023, 31, Article ID 9071969. https://doi.org/10.1155/2023/9071969.

Alramadhan, M., & Sha, K. (2017). An overview of access control mechanisms for internet of things. In 2017 *26th International Conference on Computer Communication and Networks (ICCCN)* (pp. 1–6). Vancouver, BC, Canada. https://doi.org/10.1109/ICCCN.2017.8038503.

Ameyed, D., Jaafar, F., Petrillo, et al. (2023). Quality and security frameworks for IoT architecture models evaluation. *SN Computer Sci*ence, 4, 394. https://doi.org/10.1007/s42979-023-01815-z.

Azrour, M., Mabrouki, J., Guezzaz, A., & Kanwal, A. (2021). IoT security: challenges and key issues. *Security and Communication Networks*, 2021, 11, Article ID 5533843. https://doi.org/10.1155/2021/5533843.

Bali, V., Bali, S., Gaur, D., Rani, S., & Kumar, R. (2023). Commercial-off-the shelf vendor selection: a multi-criteria decision-making approach using intuitionistic fuzzy sets and TOPSIS. *Operational Research in Engineering Sciences: Theory and Applications*, 6(2).

Beaulieu, R., Treatman-Clark, S., Shors, D., Weeks, B., Smith, J., & Wingers, L. (2015). The SIMON and SPECK lightweight block ciphers. In *Proceedings—Design Automation Conference*. https://doi.org/10.1145/2744769.2747946.

BenMbarak, O., Naanaa, A., & ElAsmi, S. (2023). New secure access control model for cloud computing based on chaotic systems. In *International Wireless Communications and Mobile Computing (IWCMC)* (pp. 1485–1490). Marrakesh, Morocco. https://doi.org/10.1109/IWCMC58020.2023.10183120.

Bhambri, P., & Rani, S. (2024). Challenges, opportunities, and the future of industrial engineering with IoT and AI. In *Integration of AI-Based Manufacturing and Industrial Engineering Systems with the Internet of Things* (pp. 1–18). USA: CRC Press.

Bhambri, P., Rani, S., Balas, V. E., & Elngar, A. A. (Eds.). (2023). *Integration of AI-Based Manufacturing and Industrial Engineering Systems with the Internet of Things*. CRC Press.

Chauhan, M., & Rani, S. (2021). Covid-19: a revolution in the field of education in India. In *Learning How to Learn Using Multimedia* (pp. 23–42). New York, NY: Springer.

Dolgov, O. S., Safoklov, B. B., & Shavelkin, D. S. (2022). Intelligent access control systems. In *International Conference on Industrial Engineering, Applications and Manufacturing (ICIEAM)* (pp. 659–663). Sochi, Russian Federation. https://doi.org/10.1109/ICIEAM54945.2022.9787125.

Duan, F., Yang, Y., Zhang, M., et al. (2022). Access control mechanism for edge nodes. In *7th IEEE International Conference on Data Science in Cyberspace (DSC)* (pp. 540–547). Guilin, China. https://doi.org/10.1109/DSC55868.2022.00081.

Elgazzar, K., Khalil, H., Alghamdi, T., Badr, A., Abdelkader, G., Elewah, A., & Buyya, R. (2022). Revisiting the internet of things: new trends, opportunities and grand challenges. *Frontiers in the Internet of Things*, 1. https://doi.org/10.3389/friot.2022.1073780.

Gautam, A. K., & Kumar, R. (2021). A comprehensive study on key management, authentication and trust management techniques in wireless sensor networks. *SN Applied Science*, 3, 50, https://doi.org/10.1007/s42452-020-04089-9.

Hamza, R., Hassan, A., Huang, T., Ke, L., & Yan, H. (2019). An efficient cryptosystem for video surveillance in the IoT environment. *Complexity,* 2019, 11, Article ID 1625678. https://doi.org/10.1155/2019/1625678.

Harbi, Y., Aliouat, Z., Refoufi, A., & Harous, S. (2021). Recent security trends in internet of things: a comprehensive survey. *IEEE Access*, 9, 113292–113314. https://doi.org/10.1109/ACCESS.2021.3103725.

Hassan, A. (2021). Lightweight cryptography for the internet of things. In *Proceedings of the Future Technologies Conference (FTC), Volume 3* (pp. 25–46). Springer. https://doi.org/10.1007/978-3-030-63092-8_52.

Henriques, M. S., & Vernekar, N. K. (2017). Using symmetric and asymmetric cryptography to secure communication between devices in IoT. In *International Conference on IoT and Application (ICIOT)* (pp. 1–4). Nagapattinam, India. https://doi.org/10.1109/ICIOTA.2017.8073643.

Hou, Y., Liu, W., Lin, H., & Wang, X. (2020). Multi-layer access control mechanism based on blockchain for mobile edge computing. In *IEEE International Conference on Parallel & Distributed Processing with Applications, Big Data & Cloud Computing, Sustainable Computing & Communications, Social Computing & Networking (ISPA/BDCloud/SocialCom/SustainCom)* (pp. 285–291). Exeter, United Kingdom. https://doi.org/10.1109/ISPA-BDCloud-SocialCom51426.2020.00061.

Hu, G., Yi, Z., Lu, L., Huang, Y., Zhai, Y., Liu, J., & Yang, B. (2021). Self-powered 5G NB-IoT system for remote monitoring applications. *Nano Energy*, 87, 106140. https://doi.org/10.1016/j.nanoen.2021.106140.

Javaid, U., Jameel, F., Javaid, U., Khan, M. T. R., & Jäntti, R. (2020). Rogue device mitigation in the internet of things: a blockchain-based access control approach. *Mobile Information Systems*, 13, Article ID 8831976. https://doi.org/10.1155/2020/8831976.

Joshi, S., Stalin, S., Shukla, P. K., Shukla, P. K., Bhatt, R., Bhadoria, R. S., & Tiwari, B. (2021). Unified authentication and access control for future mobile communication-based lightweight IoT systems using blockchain. *Wireless Communications and Mobile Computing*, 12, Article ID 8621230. https://doi.org/10.1155/2021/8621230.

Kataria, A., Agrawal, D., Rani, S., Karar, V., & Chauhan, M. (2022). Prediction of blood screening parameters for preliminary analysis using neural networks. In *Predictive Modeling in Biomedical Data Mining and Analysis* (pp. 157–169). USA: Academic Press.

Kataria, A., Puri, V., Pareek, P. K., & Rani, S. (2023, July). Human activity classification using G-XGB. In *2023 International Conference on Data Science and Network Security (ICDSNS)* (pp. 1–5). Tiptur: IEEE.

Kaur, B., Dadkhah, S., Shoeleh, F., Pinto Neto, E. C., Xiong, P., Iqbal, S., Lamontagne, P., Ray, S., & Ghorbani, A. A. (2023). IoT security dataset evolution: challenges and future directions. *Internet of Things*, 22. https://doi.org/10.1016/j.iot.2023.100780.

Khalil, U., Malik, O. A., Uddin, M., & Chen, C. L. (2022). A comparative analysis on blockchain versus centralized authentication architectures for IoT-enabled smart devices in smart cities: a comprehensive review, recent advances, and future research directions. *Sensors (Basel),* 22(14), 5168. https://doi.org/10.3390/s22145168.

Khan, A., Ahmad, A., Ahmed, M., et al. (2022). Authorization schemes for internet of things: requirements, weaknesses, future challenges and trends. *Complex & Intelligent Systems,* 8, 3919–3941. https://doi.org/10.1007/s40747-022-00765-y.

Kumar, P., Banerjee, K., Singhal, N., Kumar, A., Rani, S., Kumar, R., & Lavinia, C. A. (2022). Verifiable, secure mobile agent migration in healthcare systems using a polynomial-based threshold secret sharing scheme with a blowfish algorithm. *Sensors*, 22(22), 8620.

Kumar, S., Tiwari, P., & Zymbler, M. (2019). IoT is a revolutionary approach for future technology enhancement: a review. *Journal of Big Data*, 6, 111. https://doi.org/10.1186/s40537-019-0268-2.

Lee, D., Kim, D.-C., Kwon, D., & Kim, H. (2014). Efficient hardware implementation of the lightweight block encryption algorithm LEA. *Sensors,* 14(1), 975–994. https://doi.org/10.3390/s140100975.

Litvinov, E., Llumiguano, H., Santofimia, M. J., Del Toro, X., Villanueva, F. J., & Rocha, P. (2023). Code integrity and confidentiality: an active data approach for active and healthy aging. *Sensors (Basel),* 23(10), 4794. https://doi.org/10.3390/s23104794.

Liu, Y., Yang, X., Wen, W., & Xia, M. (2021). Smarter grid in the 5G era: a framework integrating power IoT with a cyber physical system. *Frontiers in Communications and Networks*, 2. https://doi.org/10.3389/frcmn.2021.689590.

López Delgado, J. L., Álvarez Bermejo, J. A., & López Ramos, J. A. (2022). Homomorphic asymmetric encryption applied to the analysis of IoT communications. *Sensors*, 22(20), 8022. https://doi.org/10.3390/s22208022.

Lv, Z. (2022). Practical application of IoT in the creation of intelligent services and environments. *Frontiers in the Internet of Things*, 1. https://doi.org/10.3389/friot.2022.912388.

Mathur, S., & Arora, A. (2020). IoT and PKI-based security architecture. In P. Kumar et al. (Eds.), *Industrial IoT and cyber-physical systems: transforming the conventional to digital* (pp. 25–46). IGI Global. https://doi.org/10.4018/978-1-7998-2803-7.ch002.

Mazhar, T., Talpur, D. B., Shloul, T. A., Ghadi, Y. Y., Haq, I., Ullah, I., Ouahada, K., & Hamam, H. (2023). Analysis of IoT security challenges and its solutions using artificial intelligence. *Brain Science,* 13(4), 683. https://doi.org/10.3390/brainsci13040683.

Meng, T. X., & Buchanan, W. (2020). Lightweight cryptographic algorithms on resource-constrained devices. *Preprints*, 2020, 2020090302. https://doi.org/10.20944/preprints202009.0302.v1.

Mosenia, A., & Jha, N. K. (2017). A comprehensive study of security of internet-of-things. *IEEE Transactions on Emerging Topics in Computing*, 5(4), 586–602. https://doi.org/10.1109/TETC.2016.2606384.

Pons, M., Valenzuela, E., Rodríguez, B., Nolazco-Flores, J. A., & Del-Valle-Soto, C. (2023). Utilization of 5G technologies in IoT applications: current limitations by interference and network optimization difficulties—a review. *Sensors*, 23, 3876. https://doi.org/10.3390/s23083876.

Puri, V., Kataria, A., Solanki, V. K., & Rani, S. (2022, December). AI-based botnet attack classification and detection in IoT devices. In *2022 IEEE International Conference on Machine Learning and Applied Network Technologies (ICMLANT)* (pp. 1–5). Slvador: IEEE.

Ragothaman, K., Wang, Y., Rimal, B., & Lawrence, M. (2023). Access control for IoT: a survey of existing research, dynamic policies and future directions. *Sensors (Basel),* 23(4), 1805. https://doi.org/10.3390/s23041805.

Rani, S., Bhambri, P., & Kataria, A. (2023a). Integration of IoT, big data, and cloud computing technologies. In *Big Data, Cloud Computing and IoT: Tools and Applications.*USA: CRC Press.

Rani, S., Kataria, A., & Chauhan, M. (2022). Cyber security techniques, architectures, and design. In *Holistic Approach to Quantum Cryptography in Cyber Security* (pp. 41–66). USA: CRC Press.

Rani, S., Kaur, J., & Bhambri, P. (2023b). Technology and gender violence: victimization model, consequences and measures. In *Communication Technology and Gender Violence* (pp. 1–19). Cham and New York, NY: Springer International Publishing.

Rani, S., Kumar, S., Kataria, A., & Min, H. (2023c). SmartHealth: an intelligent framework to secure IoMT service applications using machine learning. *ICT Express*, 10(2), 420–425.

Rani, S., Mishra, A. K., Kataria, A., Mallik, S., & Qin, H. (2023d). Machine learning-based optimal crop selection system in smart agriculture. *Scientific Reports*, 13(1), 15997.

Rani, S., Pareek, P. K., Kaur, J., Chauhan, M., & Bhambri, P. (2023e, February). Quantum machine learning in healthcare: developments and challenges. In *2023 IEEE International Conference on Integrated Circuits and Communication Systems (ICICACS)* (pp. 1–7). Raichur: IEEE.

Samaila, M. G., Neto, M., Fernandes, D. A. B., Freire, M. M., & Inácio, P. R. M. (2018). Challenges of securing IoT devices: a survey. *Security and Privacy,* 1(2). https://doi.org/10.1002/spy2.20.

Sankaran, K. S., Vasudevan, N., Prakash, V. R., & Guru Diderot, P. K. (2019). Access control based efficient hybrid security mechanisms for cloud storage. In *International Conference on Communication and Signal Processing (ICCSP)* (pp. 0564–0567). Chennai, India. https://doi.org/10.1109/ICCSP.2019.8698037.

Schiller, E., Aidoo, A., Fuhrer, J., Stahl, J., Ziörjen, M., & Stiller, B. (2022). Landscape of IoT security. *Computer Science Review*, 44, 100467. ISSN 1574–0137. https://doi.org/10.1016/j.cosrev.2022.100467.

Shafiq, M., Gu, Z., Cheikhrouhou, O., Alhakami, W., & Hamam, H. (2022). The rise of "internet of things": review and open research issues related to detection and prevention of IoT-based security attacks. *Wireless Communications and Mobile Computing*, 2022, 12, Article ID 8669348. https://doi.org/10.1155/2022/8669348.

Sidhu, S., Mohd, B. J., & Hayajneh, T. (2019). Hardware security in IoT devices with emphasis on hardware trojans. *Journal of Sensor and Actuator Networks*, 8(3), 42. https://doi.org/10.3390/jsan8030042.

Silva, C., Cunha, V. A., Barraca, J. P., et al. (2023). Analysis of the cryptographic algorithms in IoT communications. *Information Systems Frontiers*. https://doi.org/10.1007/s10796-023-10383-9.

Singh, H., & Rani, S. (2023). Flex sensor integrated smart strap to verify correct wearing of the face mask. *IEEE Sensors Journal*, 24(2), 2020–2027.

Singh, I., & Singh, B. (2023). Access management of IoT devices using access control mechanism and decentralized authentication: a review. *Measurement: Sensors*, 25. https://doi.org/10.1016/j.measen.2022.100591.

Sudevan, S., Barwani, B., Al Maani, E., Rani, S., & Sivaraman, A. K. (2021). Impact of blended learning during Covid-19 in Sultanate of Oman. *Annals of the Romanian Society for Cell Biology*, 14978–14987.

Taherdoost, H. (2023). Security and Internet of Things: benefits, challenges, and future perspectives. *Electronics*, 12(8), 1901. https://doi.org/10.3390/electronics12081901.

Tanwar, R., Chhabra, Y., Rattan, P., & Rani, S. (2022, September). Blockchain in IoT networks for precision agriculture. In *International Conference on Innovative Computing and Communications: Proceedings of ICICC 2022, Volume 2* (pp. 137–147). Singapore: Springer Nature Singapore.

Tariq, U., Ahmed, I., Bashir, A. K., & Shaukat, K. (2023). A critical cybersecurity analysis and future research directions for the internet of things: a comprehensive review. *Sensors (Basel)*, 23(8), 4117. https://doi.org/10.3390/s23084117.

Taurshia, A., Jaspher Willsie Kathrine, G., Souri, A., Vinodh, S. E., Vimal, S., Li, K.-C., & Sudhakar Ilango, S. (2022). Software-defined network aided lightweight group key management for resource-constrained IoT devices. *Sustainable Computing: Informatics and Systems*, 36. https://doi.org/10.1016/j.suscom.2022.100807.

Tawalbeh, L., Muheidat, F., Tawalbeh, M., & Quwaider, M. (2020). IoT privacy and security: challenges and solutions. *Applied Sciences*, 10(12), 4102. https://doi.org/10.3390/app10124102.

Ullah, I., Zeadally, S., Amin, N. U., Khan, M. A., & Khattak, H. (2021). Lightweight and provable secure cross-domain access control scheme for IoT-based wireless body area networks (WBAN). *Microprocessors and Microsystems*, 81. https://doi.org/10.1016/j.micpro.2020.103477.

Uppuluri, S., & Lakshmeeswari, G. (2023). Secure user authentication and key agreement scheme for IoT device access control based smart home communications. *Wireless Networks*, 29, 1333–1354. https://doi.org/10.1007/s11276-022-03197-1.

Williams, P., Dutta, I. K., Daoud, H., & Bayoumi, M. (2022). A survey on security in IoT with a focus on the impact of emerging technologies. *Internet of Things*, 19. ISSN 2542-6605. https://doi.org/10.1016/j.iot.2022.100564.

Yang, W., Wang, S., Yin, X., Wang, X., & Hu, J. (2022). A review on security issues and solutions of the internet of drones. *IEEE Open Journal of the Computer Society*, 3(01), 96–110. https://doi.org/10.1109/OJCS.2022.3183003.

Yu, A., Kang, J., Jiang, W., & Lin, D. (2023). FACT: a flexible access control technique for very large scale public IoT services. In *IEEE International Conference on Pervasive Computing and Communications Workshops and other Affiliated Events (PerCom Workshops)* (pp. 386–391). Atlanta, GA, USA. https://doi.org/10.1109/PerComWorkshops56833.2023.10150257.

Zhang, H., Babar, M., Tariq, M. U., Jan, M. A., Menon, V. G., & Li, X. (2020). SafeCity: toward safe and secured data management design for IoT-enabled smart city planning. *IEEE Access*, 8, 145256–145267. https://doi.org/10.1109/ACCESS.2020.3014622.

5 Introduction to Blockchain Technology

Geetabai S. Hukkeri, Shilpa Ankalaki, and Dhananjaya G.M.

5.1 INTRODUCTION

Blockchain is commonly described as a connected series of blocks that involve multiple transactions, creating a decentralized and tamper-resistant data store applicable in various domains, including electronic voting, crowdfunding, distributed resources, public record-keeping, and identity management. Traditionally, currency transactions between individuals or organizations have relied on third-party intermediaries. However, blockchain disrupts this norm by enabling direct peer-to-peer transactions. Consequently, it is possible to revolutionize the information technology landscape. Various industries, such as IoT (Panarello et al., 2018), finance, supply chain management, healthcare (Esposito et al., 2018), and reputation systems have embraced different implementations of blockchain technology, leveraging its unique strengths. Businesses seek to incorporate secure and private online transactions to enhance productivity in their operations. The utilization of information and communication technology has been pivotal in driving economic growth (Farhadi et al., 2012). Notably, the deployment of blockchain technology has attracted substantial investments, with technology companies and financial services investing over $1 billion in 2016 alone, and this amount is likely to surge in the coming years.

The appearance of blockchain technology has addressed the ongoing concerns about system security in the online realm. Here, the authors emphasize the significance of privacy, security, and trust in current electronic technologies, particularly e-commerce. The security of e-commerce encompasses critical aspects such as data security, integrity, privacy, and broader information security considerations. In conventional financial transactions, banks play a crucial role as trusted third parties (Fanning et al., 2016). Their intervention ensures reliability and trust between parties engaging in trade ventures. However, in an economy without banks, relying solely on peer-to-peer trading can lead to difficulties in establishing trust between parties. Online purchases, for example, may not guarantee genuine items, as fraud or fake product deliveries can occur. To beat this impasse, blockchain knowledge has been imagined as a solution (Casino et al., 2019). By adopting blockchain, people now have a viable alternative—a trusted intermediary that facilitates secure online transactions. This chapter emphasizes the importance of exploring by promoting trust within peer-to-peer networks. Blockchain technology plays a significant role, effectively addressing protection and confidentiality concerns prevalent in

DOI: 10.1201/9781003460367-5

the internet environment, thereby benefiting both business-oriented individuals and organizations.

The compass of this topic involves a comprehensive review of blockchain handling across various domains, together with cryptocurrency. The inception of blockchain-based cryptocurrency, Bitcoin, in 2008 marks the initiation of examples for various blockchain functions such as copyright protection, advertising, energy, healthcare, and societal applications. Further instances of blockchain functioning can be mentioned in the subsequent section. The authors marked the opening of a wave of other cryptocurrencies resembling Litecoin, Bytecoin, SwiftCoin, and Namecoin (Bonneau et al., 2015). Blockchains have been leveraged to consolidate fragmented healthcare records, enabling the tracking of personal health data. In the domain of digital advertising, MetaX utilizes blockchain technology to combat fraud. The insurance business also benefits from blockchain with platforms like InsureX; this technology-based insurance marketplace aimed at improving efficiency. Our chapter serves as a timely resource for individuals and association attracted in blockchain applications, and we believe it will inspire the exploration of even more diverse use cases. While previous review articles on blockchain have focused on technical aspects such as safety measures and agreement protocols (Wang et al., 2018), or detailed applications like the Internet of Things (Conoscenti et al., 2016) and the banking, our effort provides a broader view of applications in various sectors.

The subsequent sections are organized as follows: Section 5.2 elaborates on blockchain architecture, Section 5.3 presents various types of blockchains, while Section 5.4 compares the various blockchain networks. Section 5.5 deals with blockchain applications, and finally, Section 5.6 concludes the chapter.

5.2 BLOCKCHAIN ARCHITECTURE

Blockchain can be conceptualized as a sequential arrangement of blocks forming an open and public ledger containing the entire network's data. Each block comprises two main components. Figure 5.1 illustrates the block header, which consists of six key components (Bendiab et al., 2018).

1. *Block Version:* The block version ensures compliance with specific block validation rules.
2. *Merkle Tree Root Hash:* When a transaction occurs, data undergoes encryption using hashing algorithms, and the resulting encrypted data is distributed to every node. The blockchain employs the Merkle tree function to generate a final hash value known as the Merkle tree root, as each node's block may contain thousands of transaction records.
3. *Timestamp:* A timestamp is generated for each block, recording the current time in seconds.
4. *Difficulty Target*: The difficulty target serves as the threshold for determining a valid hashed block.
5. *Nonce:* A 4-byte field that typically starts at zero and increments with each successive hash calculation.
6. *Parent Block Hash:* This 256-bit hash value points to the previous block in the blockchain sequence.

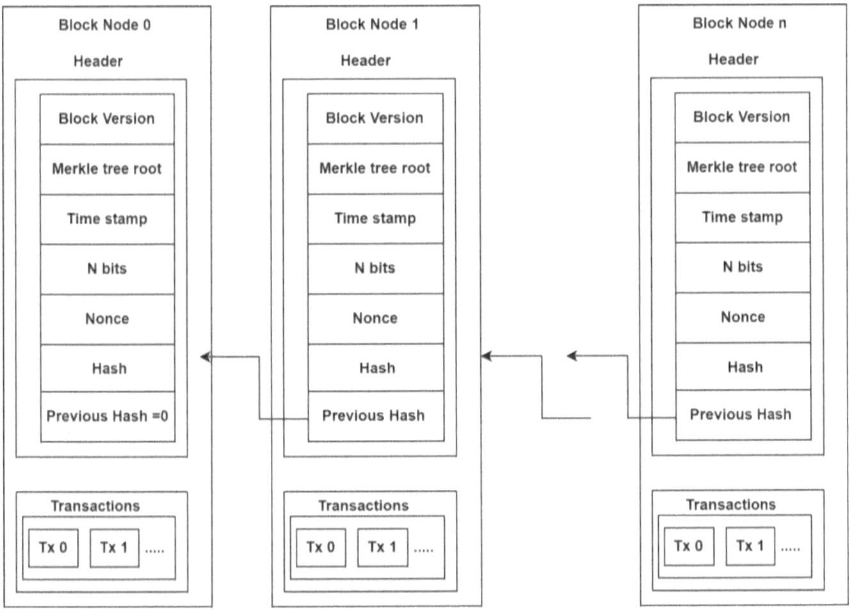

FIGURE. 5.1 The general architecture of blockchain.

Working of Blockchain: The underlying working of blockchain technology supports the concept of constructing blocks containing data and connecting them together to form a chain (Niranjanamurthy et al., 2019). Each block in the chain contains the header of the previous block, creating an unbroken and immutable sequence. If any data in a previous block is altered, the hash key associated with that block will change, resulting in a discrepancy with subsequent blocks in the chain. This tamper-resistant feature ensures the integrity of the data. When a user initiates a transaction and wishes to send data to others, this transaction is presented as a block. To add the block to the blockchain, it must be connecting with all the nodes on the network. The block is authenticated and additional to the blockchain when miners on the node approve it. Miners earn the right to approve a block by solving complex computational problems during the block's creation. Once the block is authenticated, the transaction is completed, and it becomes a separate section of the blockchain. The next new part contains decisive which user will have the privilege of publishing the next block. The group of validated blocks is linked together to form a chain, establishing the blockchain system (Zhu et al., 2019).

5.3 VARIOUS TYPES OF BLOCKCHAINS

With the increasing integration of blockchain technology into business systems, it has been classified into four main types based on its use cases:

- *Public Blockchain:* Public blockchains are open-source networks that permit anyone to participate as users, developers, network members, or miners.

There are no restrictions on joining, and equal participation is encouraged. Transactions on public blockchains are transparent and accessible to all network participants, enabling the examination of transaction details.

- *Private Blockchain:* Private blockchains are authorization networks where individuals require authorization to join. Transactions on private blockchains are private and only accessible to network participants with the necessary permissions to operate within the network.
- *Hybrid Blockchain:* Hybrid blockchains combine features from both public and private blockchains. They offer the privacy and security of private blockchains along with the transparency of public blockchains. This flexibility empowers businesses to operate with both private and publicly shared data as per their requirements.
- *Consortium Blockchain:* Consortium blockchains, also known as federated blockchains, allow new participants to join an existing blockchain structure and share data without starting from scratch. This feature helps organizations save time and development costs by quickly securing solutions for their needs.

5.4 COMPARISON OF VARIOUS BLOCKCHAIN NETWORKS

The evolution of BT over the years has progressed through various generations, each introducing an upgraded agreement instrument to address the boundaries of its predecessors. For example, third-generation blockchains have significantly improved scalability compared to first- and second-generation blockchains.

One of the notable advancements in the blockchain's generational development is the preface of smart contracts, which were lacking in the initial generation. Additionally, third-generation blockchains offer improved features such as interoperability and higher transactions per second (TPS) rates. To gain a comprehensive understanding of different blockchain platforms and their capabilities, a tabular comparison is essential. This comparison allows us to assess various blockchain protocols, their unique features, and their potential for successful implementation in our businesses. Please refer to Table 5.1 for specific comparison.

5.5 BLOCKCHAIN APPLICATIONS

This section includes discussion of various blockchain applications.

5.5.1 Cryptocurrency

Since the launch of the first cryptocurrency, Bitcoin, on the blockchain, numerous other cryptocurrencies have emerged. Bitcoin's distinct features, including anonymity, verifiability, decentralization, and consensus mechanisms, have contributed to its remarkable value, reaching $6,300 per BTC [18]. Concurrently, the economic sector has emerged as the more active and prominent adopters of blockchain, largely due to the creation of smart contracts. Beyond cryptocurrencies, the financial services industry has rapidly embraced blockchain for various applications, including

TABLE 5.1
Comparison of Various Blockchain Networks

Protocols/Parameters	XDC	Cardano	Solana	Ripple	Ethereum	Polkadot	Stellar
TPS	2005+ (approximately)	255 (approximately)	1955 (approximately)	1505 (approximately)	35 (approximately)	167 (approximately)	2505+ (approximately)
Scalability	Y (Sharding)	N	Y (horizontal PoH)	N (only by channels)	N	Parachains	Y
Smart contracts	Y	Y (KEVM)	Y (Solana BPF)	No	Y (EVM)	Parachains (EVM, Wasm)	Y
Level of decentralization	H	H	Low	M	H	H	H
On-chain governance	Y	N	N (only for programs)	Y	N	Y	Y
Human readable address	Y	N	Y	N	Y	N	Y
Interoperability	Y	N	N	N	N	Y	N
Digital identity management	Y	N	N	N	Y	N	Y
Gas fee	Nearer to zero	Nearer to zero	Nearer to zero	Nearer to zero	$3–$15 (According to network's traffic)	Nearer to zero	Nearer to zero
Decentralized exchange	Y	Y	Y	Y (Codebase)	Y	Y	Y
Deposit time	Nearer to instant	Nearer to instant	Near to instant	Nearer to instant	5 minutes	3 minutes	Nearer to instant
Data privacy	Y	N	N	N	N	N	Y
Decentralized finance	Y	Y	Y		Y	Y	Y
Currency	XDC	ADA	SOL	XRP	ETH	DOT	XLM
dApp	Y	Y	Y	N	Y	Y	Y
Chain generation	III Gen	III Gen	II Gen	I Gen	II Gen	III Gen	II Gen
Consensus mechanism	Delegated proof-of-the stake	Proof-of-the-stake	Proof-of-the-stake based on BFT (Byzantine fault tolerant)	RPCA (Ripple protocol consensus algorithm)	Proof-of-the-work	Proof-of-the-stake	Proof-of-the-stake based on federated voting
Block time	2 seconds	20 seconds	0.8 seconds	4 seconds	14 seconds	6 seconds	3–5 seconds

Y = yes, N = no, H = high, M = medium.

stock exchanges, global payments, buyback agreements, and digital identities. For instance, the Australian Securities Exchange has explored the use of Bitcoin technology to replace the existing clearing system, aiming to reduce transaction costs, enhance speed and security, and improve overall efficiency. Financial institutions like Oxygen, a London-based trading firm, have also publicized the release of their Repurchase blockchain platform (Repos). This platform allows banks and borrowers to initiate repurchase agreements by transferring funds and collateral to a predefined smart contract address. Subsequently, this smart contract safeguards the movement of collateral and deposits the funds into the borrower's account and records every transaction over the course of time. Due to the move toward electronic banking, blockchain is a fantastic fit for automating clearing houses and facilitating repurchase agreements. In recent years, as blockchain technology has grown increasingly, well-known, numerous banks have begun investing in blockchain-related projects. The online bank Fidor Bank, which is situated in Germany, has made significant advancements in the conventional banking industry's understanding of blockchain and virtual currency. In October 2013, Fidor Bank started services for exchanging euros for Bitcoins in association with the San Francisco-based Bitcoin exchange Kraken. The bank also collaborated. In February 2015, Fidor Bank collaborated with bitcoin.de to launch the P2P Bitcoin Transfer Service, leveraging Ripple Labs' payment technology to enable low-rate transfer services, which was another noteworthy endeavor. Through the internal development of three distinct systems, Citibank has also made progress in utilizing distributed blockchain technology. Citigroup said in July 2015 that one of its five core areas of focus would be blockchain due to its potential. Blockchain technology has significant disadvantages in everyday applications, even though it may provide answers for cross-border value exchange. Concerns about blockchain's security are the main ones. Contrary to traditional financial facilities, which are run by firms with proprietary software and hardware, blockchain applications are open and the system's code is shared among participants. Due to this, blockchain-based applications are more vulnerable to attacks than the old financial systems. How user privacy is protected by blockchain-based applications is discussed here. Data privacy is protected by the system operator in conventional financial models, which keep data on centralized servers. In contrast, every data on the blockchain is openly accessible and available to all users. Despite the "pseudo-anonymous" nature of blockchains, financial institutions believe this notion is too basic for conducting complex financial transactions that demand stringent confidentiality.

5.5.2 Healthcare

As new commercial applications for blockchain technology develop, the healthcare sector is rapidly expressing interest in making investments in it. Healthcare organizations find a number of blockchain-based features, as the program aims to achieve transparency, auditability, disintermediation, industry collaboration and new business models, to be very enticing. The dispersed nature of medical records because of transfers between various medical facilities is one major issue in healthcare IT. Blockchain offers a chance to provide a dependable platform for capturing healthcare data that may effectively merge highly fragmented records and enable seamless tracking of individual health records. Access to medical records, however, poses moral conundrums, hence the application must guarantee a foundation of high-integrity monitoring capacity. History-based diagnoses are frequently expensive security issues, fragmentation, and intrinsic complexity of medical solutions. This is addressed by blockchain technology, which makes it easier and less expensive to track services and money movements continuously. According to Matthews, the combination of BT with AI has the potential to address problems in the healthcare industry. However, in order to realize these encouraging ideas, technical challenges like data access and storage on the blockchain, as well as difficulties with privacy and policy, must be resolved.

5.5.3 Advertising

The blockchain presents solutions to fundamental challenges within the digital advertise supply chain, including fraud, inefficiencies, and a lack of transparency. The blockchain serves as a distributed, immutable, and transparent ledger. In 2017, universal expenses on digital advertise reached a staggering $209 billion, and over the past few years, many companies have been tirelessly trying to build blockchain-based advertising solutions. Using blockchain technology to address the fraud and transparency issues that plague digital advertising is the aim of a blockchain company by the name of MetaX. The open-source software program was developed to identify and eliminate dishonest vendors and resellers from the programmable supply chain. Ads.txt Plus, OTT advertising network Premion, a division of TEGNA, collaborates with tech company MadHive to create a blockchain-based transaction platform. Furthermore, MadHive announced MAD Network, a decentralized system that encourages all blockchain nodes engaged in the advertising value chain to be trusted. High-end digital channels and video supply chains are using blockchain technology, which are being implemented by significant MVPDs (Multichannel Video Programming Distributors). Smart contracts increase measurable KPIs and the quality of the user experience for customers by enabling tradeoffs, developers, and operators collaborate to strategize, aim, and analyze purchases comprehensively—a variety of channels, including digital, broadcast, and streaming. Marketers can now easily create consumer profiles using shared data from customers, thanks to the blockchain, which enables direct peer-to-peer interaction (Sudevan et al., 2021). Due to this, businesses are able to communicate with customers directly and utilize their data effectively. In essence, blockchain networks' decentralized, distributed, and Turing-complete characteristics substantially encourage the expansion and development of the digital advertising sector.

5.5.4 Insurance

Traditional insurance contracts often suffer from common issues leading to erroneous payments and the need for manual verification. These challenges are further compounded by the complexities involving clients, brokers, insurers, and reinsurers, as well as the risk-focused nature of insurance products. Reinsurance claims automation, data analysis through the Internet of Things (IoT), and fraud protection are crucial areas where the insurance sector needs to grow, but the use of blockchain as a distributed ledger offers great prospects. The insurance industry can integrate all relevant data, such as individual historical credit information, accident environmental data, and historical policy information, into the blockchain network, thanks to the inherent scalability of blockchain and the support of IoT. This will transform how risks are managed. Many early adopters have already embraced this change. The Blockchain Insurance Business Initiative (B3I) was founded by the top five insurance titans in October 2016 with the goal of evaluating the practicality of applications for blockchain and developing blockchain-based evidence of ideas (Bonneau et al., 2015). Notable examples of blockchain-based initiatives in the insurance industry include InsureX, which launched the first alternative insurance market to address inefficiencies in the current system. Aigang, on the other hand, is a unique blockchain-based insurance protocol that empowers individuals, groups, and developers to create their insurance prediction markets and services using smart contracts and risk-based tokenization. Aigang aims to build a self-insurance stage for manufacturers and insurance providers. The adoption of blockchain technology holds the probability to completely transform the insurance sector, making it more transparent, efficient, and secure.

5.5.5 Copyright Protection

Copyright concerns have remained a recurring issue as the internet has grown (Bali et al., 2023). Copyright infringement has been widespread, from the emergence of end-to-end file-sharing programs like Grokster and Napster to the unauthorized use of photos on the internet, various instances of digital piracy have been observed. Owners of copyrights frequently confront difficulties since their rights are routinely disregarded or attacked. Unauthorized file-sharing and unauthorized exploitation of protected content remain serious issues. However, blockchain technology has come to light as a viable remedy for these problems. Blockchain is a continuously updating, distributed, and decentralized digital ledger designed to ensure secure and transparent record-keeping and consistency across all copies of a file. It is virtually impossible to manipulate or corrupt because of its decentralized structure and lack of a single central repository. Once a record is added to the blockchain, it becomes immutable, and any changes are permanently recorded. This property makes it ideal for copyright protection, as within the context of images online, blockchain offers a public and easily verifiable digital ledger that includes the owner's data and a comprehensive transaction history. This technology presents a significant opportunity to address one of the major challenges in copyright protection, which is policing unauthorized use. Platforms like Binded have emerged to enable photographers to upload their images to a blockchain, establishing a unique cryptographic hash is used as a fingerprint for each copyright record, ensuring its authenticity and integrity. Binded helps to safeguard copyright and enables photographers to keep an eye on and stop unauthorized use across a range of internet channels, including social media sites like Instagram and Twitter, by tracking copyright records. Similar services are also provided by other blockchain-based copyright systems such as COPYTRACK.

However, copyright protection based on blockchain has several restrictions. The initial authentication process for image uploads is a huge barrier (Rani, Mishra, et al., 2023). The issue of proving ownership at the time of upload needs to be resolved. Monitoring the license usage of anonymous and untraceable consumers also poses problems that require further solutions (Rani, Pareek, et al., 2023, February).

5.5.6 Energy

Even the most fundamental energy and commodity deals include multiple actors in a game of equilibrium. Both parties in an energy trading transaction must coordinate and confirm transaction data from the deal's execution through its completion. A corporation may have to work with several counterparties, exchanges, brokers, logistical providers, banks, regulators, and price reporters during a transaction. In order to maintain smooth operations between various divisions and guarantee a complete understanding of the entire transaction process, effective coordination of the verification process is crucial not only between the two parties involved but also within the organization. With the potential to automate both internal and external corporate procedures, blockchain technology has the potential to completely change how energy transactions are structured (Kumar et al., 2022). Significant cost-savings can be made by streamlining these procedures. These savings can be made in labor costs, manual and semi-automatic costs, capital expenditures through quicker settlements, and technological efforts by cutting back on the use of numerous systems (Rani, Kumar, et al., 2023). Energy trading platforms and the incorporation of blockchain technology are the two primary categories into which energy trading applications can be divided. *Markets for exchanging energy:* Two other methods are being studied in this area right now. The first strategy involves using blockchain technology to completely restructure the existing energy infrastructure. Projects that exploit the decentralized features of blockchain to facilitate peer-to-peer Bitcoin transactions serve as examples of this (Kataria et al., 2022). However, some energy-related initiatives frequently employ decentralization without properly accounting for more efficient ways to manage complex systems. The second tactic aims to gradually enhance the current wholesale markets for electricity by employing blockchain technology. By utilizing blockchain technology, transaction verification in these markets is accelerated and made more economical. These projects aim to transform the existing electricity markets and may pave the way for the creation of new distributed energy markets that resemble "power mining machines." The power sector's growth and overall productivity might both be greatly accelerated by these large changes. *Finance for energy:* Many company initiatives have tried to employ blockchain technology with a focus on green energy, particularly through leveraging cryptocurrencies to raise money fundraising for green energy initiatives could be made simpler by using blockchain to connect with more potential investors. It is unclear, though, whether a decentralized network is required to speed up the funding process. Regarding sustainability, blockchain technology is being used in the energy sector to continuously record energy production, including information on the percentage of renewable energy sources and pollutant emissions. This makes reliable data possible and lowers the possibility of fraud and flawed decisions. Blockchain technology provides a solution in the context of electric vehicles (EVs), which currently struggle with deployment issues and a lack of adequate charging infrastructure (Singh & Rani, 2023). Blockchain technology makes it simple for private owners of charging infrastructure to offer services, removing some of the barriers to EV adoption (Rani, Kataria, et al., 2022). Additionally, it helps to achieve the goal of charge expenditure reduction through process simplification (Kataria et al., 2023, July). It is obvious that these advantages may help us get closer to a day when EVs are commonly used. By evaluating electricity requirements and putting smart contracts into place, EVs may also be utilized as batteries to stabilize the distribution of energy and control charging and discharge. The main obstacle to blockchain adoption in the energy sector right now is performance. Only three transactions can be processed per second by blockchain-based applications, with a maximum of seven transactions per second for Bitcoin (Nakamoto et al., 2009). A crucial limitation of blockchain is the requirement for critical mass (Bhambri et al., 2023). To be effective, blockchain must be adopted as a shared industrial infrastructure, which requires industry consensus on common standards. The number of people involved makes it very challenging to come to a consensus (Puri et al., 2022, December).

5.5.7 Education

The education sector is one where blockchain is only now starting to make headway. Only a small number of institutions have embraced blockchain technology in education, thus it is still in its infancy. Only 2% of higher education institutions were adopting blockchain as of 2019, according to a Gartner poll, while another 18% expected to do so in the following two years (Tanwar et al., 2022, September). Majority of the time, the institutions that have adopted blockchain utilize it to store and exchange academic transcripts and certificates. However, experts think that technology might transform education in a variety of ways, including by expanding chances for life-long learning, improving efficiency for teachers through smart contracts, and giving students control of their academic records, among other advantages. Although the opportunities are encouraging, data security, scalability, and cost are just a few of the problems preventing blockchain from being widely used in the education sector. Blockchain is anticipated to have a bigger influence in the field of education as technology develops (Rani, Bhambri, et al., 2023).

5.5.7.1 Benefits for Students

Students who save their certificates on a blockchain are able to control and manage their academic accomplishments, giving them the freedom to share them whenever and wherever they want. Universities have traditionally owned and controlled student data, thus students have had to rely on such institutions to access and share their academic records and accomplishments (Rani, Bhambri, et al., 2023).

This paradigm has glaring shortcomings. Physical documents may be misplaced or destroyed, students would have to pay a price to access them, and graduates of closed institutions might have trouble finding a source to vouch for their academic credentials. ITT Technical Institute, a for-profit organization, abruptly shuttered 130 sites in 38 states in 2016, leaving students and alumni without access to their data until the Department of Education intervened (Chauhan & Rani, 2021).

Blockchain enables students to take control of their academic identities by giving them ownership of their personal information. For graduates looking for work, for instance, this makes verifying the integrity of the credentials listed on their résumés much easier and offers them greater control over what an employer can access.

5.5.7.2 Benefits for Institutions

Higher education institutions can save time and money by streamlining the verification process by using blockchain to award diplomas (Bhambri & Rani, 2024). According to recent research by the University of Rome, the process of authenticating credentials costs the institution over 19,000 euros yearly, or more than $20,000, which equates to almost 36 weeks of work. Blockchain-issued diplomas also make it considerably simpler for graduate schools to validate a student's academic record because they are essentially tamper-proof (Rani, Kaur, et al., 2023).

The education sector could be profoundly impacted by blockchain technology in a number of ways. Here are some of the major spheres in which blockchain can contribute:

a. Academic credentialing and certifications can be issued and verified using tamper-proof, decentralized systems that can be created using blockchain technology. This can guard against fake credentials and guarantee the veracity of a student's accomplishments.
b. *Student Data Management:* Blockchain provides a transparent and secure means to handle student information, including attendance, academic performance, and other pertinent data. By doing this, you can be sure that student data is securely and privately maintained and made available to those who need it.
c. *Online Learning and Content Exchange:* Blockchain can enable peer-to-peer content exchange and microtransactions, allowing teachers to share or sell their course materials with students directly. This has the potential to empower educators and content producers, particularly when it comes to online learning systems.
d. *Copyright Protection:* The immutability and timestamping capabilities of blockchain technology can aid in the protection of intellectual property rights by ensuring that educational content, research papers, and other materials are appropriately credited to their authors and are not vulnerable to unauthorized revisions.
e. *Decentralized Learning Platforms:* Blockchain can help with the creation of decentralized learning environments that reward student and educator engagement and cooperation. Students may be encouraged to participate actively in the learning process by awarding contributions with tokens or other digital assets.
f. *Micro-credentials and Lifelong Learning:* Blockchain can help with the development of badges and micro-credentials for more compact, specialized abilities or competencies. This enables people to demonstrate their knowledge and ongoing learning efforts outside of the context of traditional academic degrees.

g. *Openness and Accountability:* Blockchain-based solutions at educational institutions can increase openness and accountability in all administrative procedures, including resource allocation, financial transactions, and decision-making.
h. *Global Collaboration and Recognition:* Blockchain can promote global partnerships and make it easier to recognize credentials obtained in other nations, potentially lowering obstacles to international education and employment mobility.
i. *Smart Contracts for Education Agreements:* Smart contracts, which are self-executing contracts with predetermined conditions, can be used to automate and enforce agreements between students, educators, and educational institutions. This streamlines procedures like fee payments, enrollment, and other related activities.
j. *Data Analytics and Research:* Blockchain can promote data collaboration and sharing while maintaining data integrity. Institutions of higher learning can safely pool data to acquire insights into student performance, learning trends, and pedagogical efficacy.

However, there remain obstacles to be solved, such as scalability, interoperability, and regulatory concerns, despite the tremendous potential of blockchain technology in education. We may anticipate additional cutting-edge applications and broad adoption in the future as the technology develops and the education industry gains experience using blockchain solutions.

5.5.8 Society Applications

Digital contracts are a cutting-edge technology that represents the next generation of networks. Architecture can transform traditional financial lending by eliminating credit concerns. Traditional loan agreements involve the lender taking on risk in addition to providing the money, which leads to high loan interest rates and frequently higher-than-loan-amount collateral requirements. Borrowers are now able to employ virtual assets instead of physical ones as collateral, eliminating the need for physical collateral and lower loan money. It is unneeded for the process of presenting credit history, employment history, or other extensive documentation. Property ownership is securely encoded in the blockchain for transparent access. *Example:* Smart contract ideas are applied to physical objects like cars and telephones using encryption technologies to secure ownership. For instance, the car key, equipped with an anti-theft device, functions only when the correct protocol is employed. A smartphone operates only with the proper password in a manner similar to this. The distributed ledger of the blockchain eliminates the physical key storage and transferability issues by permitting blockchain miners to replace and duplicate lost protocols. *Sound of Bitcoin:* Whether it be in the vinyl or digital music eras, music publishers have long faced copyright difficulties. But there is a solution: a traceable music copyright database constructed using blockchain and smart contracts. This database delivers real-time revenue allocation to the copyright owner and the performer based on customer behavior. Music enthusiasts also have the choice to pay using digital currency. On the blockchain, during the 2016 presidential election, both Democrats and Republicans expressed apprehensions regarding the fairness of the voting process. Blockchain and smart contracts provide a transparent method, enabling each voter to view their vote as well as the broader statistical process. Furthermore, blockchain technology can streamline the government's expensive but necessary process of money verification. By providing a self-managed platform, blockchain enables businesses, organizations, political entities, and citizens to interact with greater accountability. Additionally, it enables individuals to ensure that their will is appropriately conveyed through blockchain technology.

5.6 CONCLUSION

Despite the widespread deployment of blockchain applications, there remain several unresolved issues that need addressing to enhance scalability, efficiency, and overall durability. While the individual features offered by blockchains may not be unique on their own and popular mechanisms, it is the union of these attributes that construct them highly suitable for various requests, leading to significant heed from different industries.

As blockchains grow, they are looking to find applications in even more industries and domains beyond those covered in the current survey. However, it is crucial to recognize that blockchains are not a universal solution or a complete alternative to traditional databases. There are situations where conventional databases remain more appropriate and efficient. Through the survey, specific characteristics required for each application domain have been identified. This identification aids in selecting the appropriate blockchain and implementing corresponding mechanisms to convert to meet unique required of a particular request. This study provides a thorough overview of the use of blockchains in a variety of industries, including cryptocurrencies. The chapter offers a pertinent overview for organizations seeking to understand blockchain technology's potential applications in healthcare, advertising, insurance, copyright protection, energy, and societal contexts. Furthermore, it serves as a catalyst for exploring further possibilities of blockchain implementation in various other fields.

REFERENCES

Bali, V., Bali, S., Gaur, D., Rani, S., & Kumar, R. (2023). Commercial-off-the shelf vendor selection: A multi-criteria decision-making approach using intuitionistic fuzzy sets and TOPSIS. *Operational Research in Engineering Sciences: Theory and Applications*, *6*(2).

Bendiab, K., Kolokotronis, N., Shiaeles, S., & S. Boucherkha. (2018). WiP: A novel blockchain-based trust model for cloud identity management. In *IEEE 16th Int. Conf. Dependable, Autonomic Secure Comput., 16th Int. Conf. Pervasive Intell. Comput., 4th Int. Conf Big Data Intell. Comput. Cyber Sci. Technol. Congr.*) (pp. 724–729).

Bhambri, P., & Rani, S. (2024). Challenges, opportunities, and the future of industrial engineering with IoT and AI. In *Integration of AI-Based Manufacturing and Industrial Engineering Systems with the Internet of Things* (pp. 1–18). USA: CRC Press.

Bhambri, P., Rani, S., Balas, V. E., & Elngar, A. A. (Eds.). (2023). *Integration of AI-Based Manufacturing and Industrial Engineering Systems with the Internet of Things*. USA: CRC Press.

Bonneau, J., Miller, A., Clark, J., Narayanan, A., Kroll, J. A., & Felten, E. W. (2015). SoK: Research perspectives and challenges for bitcoin and cryptocurrencies. In *IEEE Symposium on Security and Privacy (SP)* (pp. 104–121). Poland: IEEE.

Casino, F., Dasaklis, T. K., & Patsakis, C. (2019). A systematic literature review of blockchain-based applications: Current status, classification and open issues. *Telematics Inform, 36*, 55–81.

Chauhan, M., & Rani, S. (2021). Covid-19: A revolution in the field of education in India. In *Learning How to Learn Using Multimedia* (pp. 23–42). New York, NY: Springer.

Conoscenti, M., Vetro, A., & De Martin, J. C. (2016). Blockchain for the internet of things: A systematic literature review. In *IEEE/ACS International Conference of Computer Systems and Applications (AICCSA)* (pp. 1–6). China: IEEE.

Esposito, C., De Santis, A., Tortora, G., Chang, H., & Choo, K.-K. R. (2018). Blockchain: A panacea for healthcare cloud-based data security and privacy? *IEEE Cloud Computing, 5*(1), 31–37.

Farhadi, M., Ismail, R., & Fooladi, M. (2012). Information and communication technology use and economic growth. *PLOS ONE, 7*(11) e48903, 1–7.

Kataria, A., Agrawal, D., Rani, S., Karar, V., & Chauhan, M. (2022). Prediction of blood screening parameters for preliminary analysis using neural networks. In *Predictive Modeling in Biomedical Data Mining and Analysis* (pp. 157–169). USA: Academic Press.

Kataria, A., Puri, V., Pareek, P. K., & Rani, S. (2023, July). Human activity classification using G-XGB. In *2023 International Conference on Data Science and Network Security (ICDSNS)* (pp. 1–5). Tiptur: IEEE.

Kumar, P., Banerjee, K., Singhal, N., Kumar, A., Rani, S., Kumar, R., & Lavinia, C. A. (2022). Verifiable, secure mobile agent migration in healthcare systems using a polynomial-based threshold secret sharing scheme with a blowfish algorithm. *Sensors, 22*(22), 8620.

Niranjanamurthy, M., Nithya, B. N., & Jagannatha, S. (2019). Analysis of blockchain technology: Pros, cons and SWOT. *Cluster Computing, 22*(6), 14743–14757.

Panarello, A., Tapas, N., Merlino, G., Longo, F., & Puliafito, A. (2018). Blockchain and Iot integration: A systematic survey, *Sensors, 18*(8), 2575, 1–37.

Puri, V., Kataria, A., Solanki, V. K., & Rani, S. (2022, December). AI-based botnet attack classification and detection in IoT devices. In *2022 IEEE International Conference on Machine Learning and Applied Network Technologies (ICMLANT)* (pp. 1–5). Salvodar: IEEE.

Rani, S., Bhambri, P., & Kataria, A. (2023). Integration of IoT, big data, and cloud computing technologies. In *Big Data, Cloud Computing and IoT: Tools and Applications*. USA: CRC Press.

Rani, S., Kataria, A., & Chauhan, M. (2022). Cyber security techniques, architectures, and design. In *Holistic Approach to Quantum Cryptography in Cyber Security* (pp. 41–66). USA: CRC Press.

Rani, S., Kaur, J., & Bhambri, P. (2023). Technology and gender violence: Victimization model, consequences and measures. In *Communication Technology and Gender Violence* (pp. 1–19). New York, NY: Springer International Publishing.

Rani, S., Kumar, S., Kataria, A., & Min, H. (2023). SmartHealth: An intelligent framework to secure IoMT service applications using machine learning. *ICT Express, 10*(2), 420–425.

Rani, S., Mishra, A. K., Kataria, A., Mallik, S., & Qin, H. (2023). Machine learning-based optimal crop selection system in smart agriculture. *Scientific Reports, 13*(1), 15997.

Rani, S., Pareek, P. K., Kaur, J., Chauhan, M., & Bhambri, P. (2023, February). Quantum machine learning in healthcare: Developments and challenges. In *2023 IEEE International Conference on Integrated Circuits and Communication Systems (ICICACS)* (pp. 1–7). Raichur: IEEE.

Singh, H., & Rani, S. (2023). Flex sensor integrated smart strap to verify correct wearing of the face mask. *IEEE Sensors Journal, 24*(2), 2020–2027.

Sudevan, S., Barwani, B., Al Maani, E., Rani, S., & Sivaraman, A. K. (2021). Impact of blended learning during Covid-19 in Sultanate of Oman. *Annals of the Romanian Society for Cell Biology*, 14978–14987.

Tanwar, R., Chhabra, Y., Rattan, P., & Rani, S. (2022, September). Blockchain in IoT networks for precision agriculture. In *International Conference on Innovative Computing and Communications: Proceedings of ICICC 2022, Volume 2* (pp. 137–147). New York, NY: Springer Nature Singapore.

Wang, W., Hoang, D. T., Xiong, Z., Niyato, D., Wang, P., Hu, P., & Wen, Y. (2018). A survey on consensus mechanisms and mining management in blockchain networks. *arXiv preprint arXiv:1805.02707*. https://arxiv.org/abs/1805.02707

Zhu, L., Gai, K., & Li, M. (2019). Blockchain and the internet of things. In *Blockchain Technology in Internet of Things* (pp. 9–28). Cham, Switzerland: Springer.

6 Trends and Prospectus of Blockchain Technology

Vaneeta M., Sangeetha V., Mamatha A., and Anita Kanavalli

6.1 INTRODUCTION

Blockchain is basically a decentralized database system that stores a constantly updating list of data that is verified by the people who run the system. This data is kept in a public record, which includes details of every transaction that has taken place. Due to its decentralized nature, blockchain eliminates the need for any intermediaries in transactions. The transaction data for each node in the network is shared and available to all nodes in the network. Due to this, the transparency of the network is much higher than that of a centralized transaction with a third party.

The anonymity of all nodes within the network enhances transaction authentication safety, with cryptography serving as a pivotal security technique in the blockchain. This method encompasses two core elements—cryptography, utilized for encrypting messages in a peer-to-peer (P2P) network, and hashing, employed to fortify block information and establish links between blocks in the blockchain. The primary focus of cryptography is to guarantee the security of users, transactions, and double-spending safeguards. It is also used to secure various transactions on a blockchain network, ensuring that only the intended recipients of the transaction data can access, read, and process the transaction.

Blockchain is mainly used in finance and cryptocurrencies. Beyond its original application, blockchain has found utility in diverse sectors like healthcare, education, and supply chain management. Its adaptability extends to asset tracking, energy management, smart home/city systems, and the Internet of Things. Figure 6.1 depicts types of blockchain.

6.1.1 Public Blockchain

A public blockchain functions as a decentralized and permissionless ledger system, enabling anyone to join and engage in transactions. This non-permissioned ledger structure ensures that each peer possesses a replicated set of information, mitigating concerns related to centralization, including security and transparency issues. Information is not stored centrally but is distributed across a peer-to-peer network, addressing the limitations associated with centralized systems. The decentralized nature necessitates a verification mechanism for data authenticity, often implemented

DOI: 10.1201/9781003460367-6

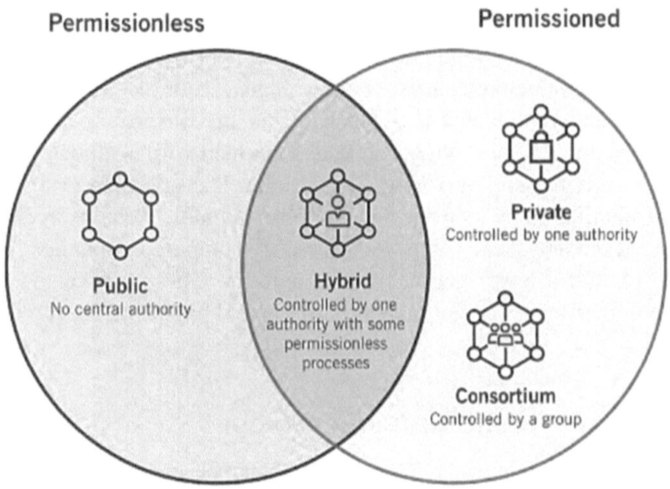

FIGURE 6.1 Types of blockchain.

through a consensus algorithm like Proof of Work (PoW) or Proof of Stake (PoS). Public blockchains provide advantages such as trust, security, openness, and transparency, exemplified in applications like voting and fundraising.

6.1.2 Private Blockchain

In the realm of blockchain networks, a private blockchain functions within a secure, confined environment and is typically under the management of a single entity. These networks operate on a restricted scale, often confined within a specific company or organization, rather than being open to the broader public. Alternately known as permissioned blockchain or enterprise blockchain, private blockchains present several advantages, including heightened speed and scalability. Furthermore, the controlling organization has the ability to establish access levels, security protocols, authorizations, and overall accessibility (Bali et al., 2023).

6.1.3 Hybrid Blockchain

Organizations aiming to strike a balance between privacy and the advantages of public blockchain technology often embrace a hybrid blockchain model. This innovative approach blends features from both public and private blockchains, empowering businesses to establish a private system with controlled access to specific blockchain data. In this hybrid setup, organizations can determine who has permission to access the data and which information is made accessible. The benefits of a hybrid blockchain encompass enhanced security, cost-effectiveness, and a measured level of transparency. This adaptable blockchain model finds applications in diverse fields, ranging from real estate to rental services.

6.1.4 Consortium Blockchain

Consortium blockchains share characteristics with hybrid blockchains, encompassing both private and public attributes, yet they engage multiple organizations within a decentralized network. Consensus mechanisms in consortium blockchains are overseen by predefined nodes, with each node equipped with a validator responsible for initiating, receiving, and verifying transactions. The initiation or acceptance of transactions is facilitated by member nodes. Consortium blockchains, recognized for their heightened security, scalability, and efficiency compared to public blockchains, do, however, exhibit a lower degree of transparency. These blockchain structures find application in diverse sectors such as banking and payments, research, and food tracking.

6.1.5 Development of Blockchain Technology

Blockchain technology has evolved incrementally since its inception in 1991. This timeline can be divided into four broad categories: early beginnings, mid struggle, the stage of acceptance, and move business applications to blockchain. As shown in Figure 6.2, stages are also referred to as "Blockchain 1.0": Bitcoin, which was designed with a single purpose to support cryptocurrency and was slow and difficult to use. "Blockchain 2.0": Ethereum, which was developed after Bitcoin. It was designed to be more than just a cryptocurrency and to facilitate a variety of decentralized applications. Finally, "Blockchain 4.0": This stage will enable businesses to move their current operations to decentralized, self-regulatory, trustless, encrypted ledgers, bringing the advantages of blockchain to the forefront.

This chapter offers an in-depth exploration of current trends and future possibilities in blockchain technology, mapping the year-by-year progression across the four stages of its development. The chapter is structured as follows: Section 6.2 delves into the foundational aspects of Blockchain 1.0. Section 6.3 explores the evolution of Blockchain 2.0. Section 6.4 examines the diverse applications of Blockchain 3.0 across various sectors. Section 6.5 highlights the substantial enhancements characterizing Blockchain 4.0. Section 6.6 provides an overview of anticipated blockchain trends shaping the business landscape in 2023, Section 6.7 emphasize the future prospects, and Section 6.8 concludes the chapter.

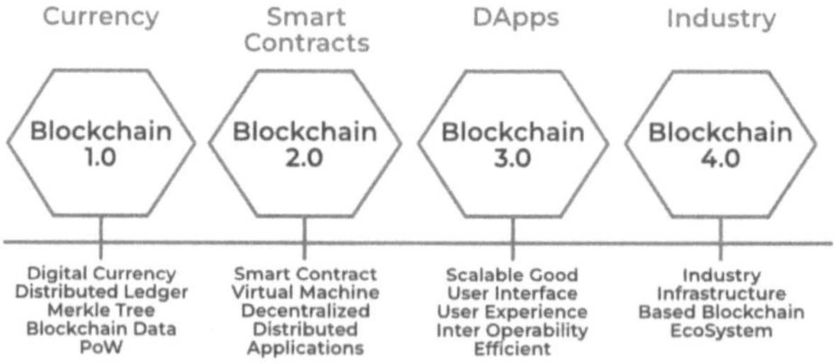

FIGURE 6.2 Phases of blockchain.

6.2 BLOCKCHAIN 1.0

This section describes the related articles published in a crisp manner. Blockchain 1.0, also known as "Blockchain" or "blockchain," is the first generation of blockchain technology developed by Satoshi Nakamoto in 2009. In late 2008, an unidentified author, under the pseudonym Satoshi Nakamoto, released a whitepaper "Bitcoin: A Peer-to-Peer Electronic Cash System." This paper introduced a consensus mechanism based on Proof of Work (PoW), leveraging software-based computing capabilities. The proposed system involves adding a timestamp and a nonce to transactions, generating a hash value, with the calculation dependent on the previous block's hash value. This process creates an unalterable chain of Proof-of-Work records based on hashing, requiring the entire Proof of Work to be redone to make any modifications. The validation of this record is ensured by other decentralized nodes within the network.

The first application of this generation of blockchain technology was Bitcoin, which was initially intended for the purpose of monitoring financial transactions, such as the transfer of value between two parties in a secure, unalterable ledger, without the need for intermediaries such as banks or governments. This version of blockchain technology is characterized by its decentralized architecture, its cryptographic security, and its capacity to facilitate peer-to-peer transactions.

6.2.1 A Brief History of Blockchain

The era from 1991 to 2008 is called early inception stage. The blockchain technology was initiated by Stuart Haber and W. Scott Stornetta working as research scientist in 1991. The aim was to develop practical solution to secure and backup the digital documents by placing a timestamp on document and create a secure chain of blocks. This chain is cryptographically secure to protect the integrity of past information. Introduced in 1992, the Merkle tree concept serves as a data structure crucial for efficiently and securely encrypting transactions. This method employs cryptographic hash functions and public-key cryptography to encrypt and verify transactions across all nodes in the network. The construction of a Merkle tree involves consolidating all transactions in a block and generating a digital signature for the entire set of operations, as illustrated in Figure 6.3.

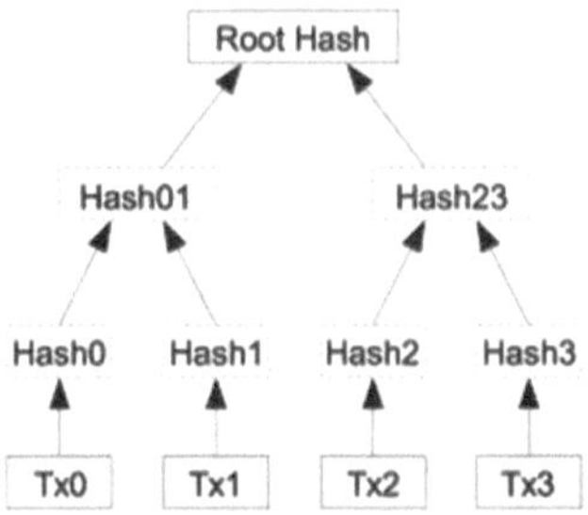

FIGURE 6.3 Merkle tree.

6.2.2 Evolutionary Trends in Blockchain 1.0

The era from 1991 to 2008 is called early inception stage. The analysis on strength, weakness, opportunities, and threats of Bitcoin was performed to look at some recent events and trends that could affect its economic paradigm. The price of Bitcoin currently stands at US$29.69 per BTC, with a circulating supply of US$19,455,281. This implies that the total market cap of Bitcoin is US$572.088 million. The volume of Bitcoin traded has decreased by US$2,446.859.564.76 over the past 24 hours, representing a decrease of 37.72%.

Bitcoin was established on January 3, 2009, with the mining of the inaugural Bitcoin block known as the "Genesis Block," carried out by Satoshi Nakamoto, who was rewarded with 50 Bitcoins. A significant milestone occurred on January 12, 2009, with the first transaction of 10 Bitcoins from Satoshi to Hal Finney. Subsequently, on October 31, 2009, the public introduction of the Bitcoin exchange marketplace, facilitating the exchange of Bitcoins for real money, marked a crucial development. Nakamoto further contributed to the dissemination of news and information by initiating the Bitcoin Talk Forum. Notably, Bitcoin's value experienced rapid growth, reaching $250 by 2012.

A new cryptocurrency called Litecoin that competed with Bitcoin was developed. The main purpose of designing Litecoin was to handle transactions of lesser value with increased processing speed. Litecoin was introduced in 2011 by Charles Lee, the founder of the cryptocurrency (Kumar et al., 2022).

A comparative analysis of two cryptocurrencies, Bitcoin (BTC) and Litecoin (LTC), was conducted by researchers Mustafa and Sulaiman in 2021 to gain insights into their recent trends and stability.

6.2.3 Advantages and Limitations of Blockchain 1.0

The use of traditional sensors in the Blockchain 1.0 was to develop a transparent, open, fully decentralized, unchangeable, and distributed ledger and transaction system for the global financial system. Some of its drawbacks are as follows:

- It is slow, processing only a few transactions per second.
- It is expensive, as it relies on miners to verify transactions.

6.3 BLOCKCHAIN 2.0

The introduction of a new version of blockchain is due to the fact that the mining of bitcoin in version 1.0 was inefficient and lack of network scalability. Consequently, the problem is addressed in phase 2.0, which allows for expansion of the blockchain beyond cryptocurrencies and include smart contracts. The second phase of blockchain technology is often termed as Blockchain 2.0 or "smart contracts." This version was created in 2015 by Ethereum's Vitalik Buterin to extend the functionality of blockchain beyond just financial transactions. The main difference between this and other versions is that it enables developers to create autonomous "smart" contracts that can store data on a distributed ledger without relying on third parties for enforcement. These contracts can be self-enforcing, reducing the need for costly dispute resolution processes such as those offered by courts and arbitration services.

6.3.1 Evolutionary Trend of Blockchain 2.0

This blockchain version indicates the middle stage evolution from 2013 to 2018. There were many rises and struggles as a future technology. The year-wise trend is depicted in following context.

At the beginning of 2013, Bitcoin revolutionized the economic landscape, with 11 million Bitcoins being traded at a value higher than $1 billion. At the same time, Thailand and China both prohibited the use of Bitcoin, leading to a resurgence of faith in the digital currency, with its value reaching a peak of $1,164 in February 2014. In 2014, Buterin established the Ethereum Foundation, a decentralized computing platform powered by a blockchain. Smart contracts, introduced in 2015, have emerged as a key application of blockchain technology. These self-executing codes operate autonomously, triggering predefined actions on the Ethereum network when specific conditions are met. The Linux Foundation later developed Hyperledger, an open-source framework designed to facilitate the creation of decentralized applications by providing the necessary frameworks, tools, and libraries.

By 2016, there was a growing recognition of the potential of blockchain technology beyond its cryptocurrency applications. IBM disclosed a cutting-edge blockchain strategy to incorporate its cloud-powered solution into BaaS offerings. Additionally, R3 announced a new distributed technology known as Corda. In 2017, the value of Bitcoin reached a peak of almost $20,000, as US regulatory authorities approved the cryptocurrency as a viable investment. The initial period of 2018 proved to be a difficult period for those who are passionate about blockchain technology.

6.3.2 Overview on Smart Contracts, Ethereum, and Hyperledger Fabric

Smart Contract: Smart contracts are simple code snippet stored on blockchain. They are executed when predetermined conditions are satisfied. This mechanism automates the execution of an agreement. There is no involvement of any intermediary and there is no time loss. Thus, it can also support to automate an entire workflow, activating the next action when conditions are satisfied.

The working process of smart contracts is outlined in Figure 6.4. Initially, the parties involved in the contract collaboratively establish the terms. Once the terms and conditions are agreed upon, they are translated into a programming code known as smart contracts. Essentially, these smart contracts encapsulate various conditional statements that outline potential scenarios for future transactions. Subsequently, these smart contracts are stored on the blockchain network and replicated among the participants within the blockchain. The code is then executed by all participants in the network. If the contractual terms are satisfied, verified by all blockchain participants, the corresponding transaction is executed. Assets are released to participants based on predefined policies. Upon completion of the transaction, all pertinent details are recorded on the blockchain.

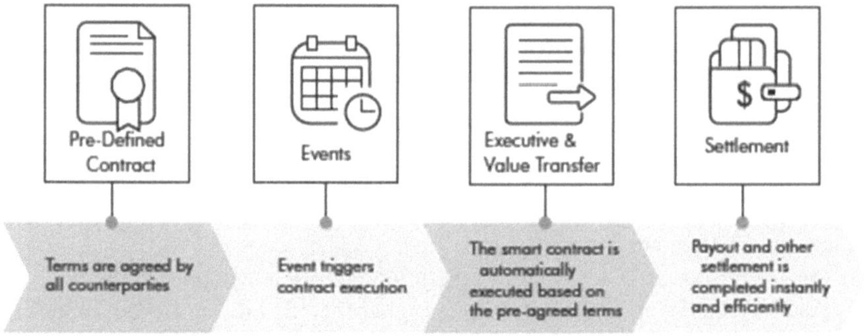

FIGURE 6.4 Process in smart contract.

Many authors conducted systematic mapping studies encompassing all peer-reviewed research in smart contracts within the technology domain. Several in-depth reviews of the smart contract-based blockchain literature were published during 2012–2022.

Ethereum: The evolution of smart contracts necessitated the creation of a platform for their verification and execution, leading to the development of Ethereum. Ethereum stands out as a decentralized blockchain platform designed to securely execute and verify smart contracts within a peer-to-peer network. Ether, the native currency of Ethereum, plays a central role in this ecosystem. Ethereum accounts are established to facilitate the sending and receiving of transactions (Rani, Kaur, et al., 2023). When conducting transactions, a sender is required to sign them and expend Ether, serving as the transaction processing cost on the network. Collaborative framework based on blockchain technology that utilizes smart contracts for resource sharing was proposed by Agrawal et al. (2023).

An application for auction using optimized smart contract was introduced by Huang et al. (2021). The paper explored the structure of Ethereum and security problems on smart contract. Proof-of-concept implementation of the Ethereum blockchain, Proof of Authority, on an Internet of Things (IoT) system using a real-world application was proposed by Alrubei et al. (2019). An overview of the Ripple protocol, security and privacy measures associated with the Bitcoin protocol, and Ripple consensus protocol were presented by Armknecht et al. (2015).

Hyperledger Fabric: In 2016, the Linux Foundation introduced Hyperledger, a framework aimed at fostering the development of blockchain-based distributed ledgers. Hyperledger is a comprehensive ensemble comprising open-source frameworks, standards, tools, and libraries, specifically crafted for the construction of blockchains and associated applications (Chauhan & Rani, 2021).

Li et al. (2020) have presented the Hyperledger greenhouse, featuring various frameworks and tools strategically designed to facilitate the development of enterprise blockchain. In various studies, demonstration of the applications of Hyperledger Fabric in managing IoT data were showcased. Meanwhile, Honar Pajooh et al. (2021) proposed an integrated IoT system employing Hyperledger Fabric to safeguard edge computing devices through a local authentication mechanism. Furthermore, reviews

FIGURE 6.5 Decentralized applications architecture.

were conducted on comprehensive evaluation of Hyperledger Fabric's performance (Bhambri & Rani, 2024).

6.3.3 Advantages and Limitations of Blockchain 2.0

Blockchain 2.0 successfully processed high number of transactions on public network and allowed the development of decentralized applications (DApps) (Kataria et al., 2023, July). It is much faster due to its smart contract capabilities. It is relatively expensive since they rely on miners to validate transactions (Puri et al., 2022, December).

6.4 BLOCKCHAIN 3.0

This marks the intermediate stage in the evolution of blockchain technology. The initial iterations, Blockchain 1.0 and 2.0, primarily relied on Proof of Work (PoW), which had the drawback of taking hours to validate transactions, rendering it non-scalable and contributing to its ultimate failure. The third phase of blockchain development is characterized by the emergence of decentralized applications (dApps). These digital programs operate on a network of computers forming a blockchain, eliminating reliance on a single centralized authority. Blockchain 3.0 incorporates advanced consensus methods such as Proof of Stake (PoS) and Proof of Authority (PoA). These consensus mechanisms are instrumental in accelerating the speed and computational power of smart contracts, all achieved without the necessity for separate transaction fees. Figure 6.5 shows the architecture of decentralized applications encompasses elements such as the back end, security measures, and design patterns. The DApps are composed of a front-end code that interacts with the user, as well as a backend code (smart contract) that operates on a blockchain or decentralized network, where it is not subject to manipulation or control by a single entity (Rani, Kumar, et al., 2023).

Applications like YouTube, Facebook, and Instagram are centralized, meaning that they are owned and managed by a single authority. They run on a central server that takes in user requests, handles them, and gives or denies access based on the authority's rules. Decentralized applications, on the other hand, are not owned by anyone or a company. They are deployed on distributed systems, meaning there is no single point of failure since individual users' computers do not rely on a central server to manage processes. Examples of decentralized apps include BitTorrent and Augur, Golem and Melonport, and OMG Network.

6.4.1 Evolutionary Trends in Blockchain 3.0

Private blockchains are not decentralized, but rather operate as a distributed database that is secured with cryptographic principles and the requirements of the organization (Singh & Rani, 2023). Only authorized parties are permitted to run a complete node, execute transactions, or verify/authenticate changes to the blockchain (Kataria et al., 2022). These blockchains are widely used by businesses in a variety of industries, including retail and healthcare, insurance, and financial services, as well as in government. They offer robust, fast, highly secure, and scalable database services to organizations. Following are the research publications on dApps, Private, hybrid blockchain, and usage of blockchain in various applications (Rani et al., 2022).

An overview of the current state of blockchain technology, including the introduction of decentralized applications (dApps) as a new form of blockchain-powered software, was presented by Cai et al. (2018). Ethereum has enabled the development of various categories of decentralized applications, such as exchange, energy, and energy-related services. Decentralized trust model is developed to secure data, transactions, and smart contracts by Ncube et al. (2020). It includes a privacy-preserving messaging protocol and two smart contracts—PPP to set up an attribute-based trust model and PSCSM to provide secure data access control using the ABE scheme. A solution to the issue of double energy expenditure in Micro-Grid, which poses a risk to the security of trading infrastructure, is proposed. Performance and limitations of two of the most widely used private blockchain platforms—Hyperledger Fabric (private deployment) and Ethereum (public deployment)—are elaborated by Pongnumkul et al. (2017). The results of an experimental analysis, based on a variety of transactions, demonstrate that Hyperledger Fabric performs better than Ethereum with respect to evaluation metrics of execution time, latency, and throughput. Blockchain has also been used in other domains and industries. EduCTX is a worldwide higher education blockchain-based credit system. It is based on the concept of the European Credit Transfer and Accumulation System proposed by Turkanović et al. (2018). A reliable system of academic credit can be established using blockchain technology. The first application of a novel platform for tracking learning accomplishments outside of transcripts and certificates was proposed by Ocheja et al. (2019).

The extensive global disruption caused by COVID-19, impacting millions of individuals, prompted the strict monitoring of people to ensure compliance with social distancing measures. Digital contact tracing emerged as an effective technique to identify and isolate affected individuals. Idrees et al. (2021) proposed the utilization of blockchain technology to implement this technique, emphasizing its efficacy in safeguarding users' personal information and ensuring privacy and security. Addressing the healthcare and telemedical laboratory services sector, many authors suggested research integrating blockchain, IoT, and cloud technologies in the medical environment (Sudevan et al., 2021). Examining the intersection of Electronic Health Records (EHR) and blockchain technology, Mayer et al. (2020) conducted a systematic literature analysis highlighting previous research. The findings underscored blockchain's potential to enhance the privacy, security, and interoperability of health data. Shifting focus to the insurance sector, several applications of the blockchain technology were developed. In the realm of IoT and blockchain integration for efficient routing, Awan et al. (2020) introduced a novel approach by devising a cluster-based multipath routing system. Addressing air quality monitoring, Benedict et al. (2019) demonstrated the utilization of the blockchain network in an Air Quality Monitoring System using gas readings of SO_2, NO_2, RSPM/PM10, and PM2.5 in a smart city, storing them as untampered blocks. In a separate initiative, Vaneeta et al. (2023) introduced a tamper-proof air quality management system by incorporating machine learning model algorithms. The considered air pollutants in this context were PM2.5, PM10, CO, SO_2, and NO_2.

Westerlund et al. (2021) propose a study with the goal of comprehending the advantages of integrating blockchain technology into the food supply chain (Rani, Mishra, et al., 2023).

6.5 BLOCKCHAIN 4.0

The propitious evolution in the blockchain development is the Blockchain phase 4.0. Blockchain 4.0 is all about innovation. The top three priorities are speed, getting people to use it, and getting it to a wider range of people. The phase is called the Global Acceptance of Blockchain as a Potential Technology. It covers the timeline of Blockchain from 2019 to the Present.

6.5.1 Evolutionary Trends in Blockchain 4.0

In 2019, there was an increase in the number of people buying Bitcoin and investing in cryptocurrencies, like stock market and mutual fund investments. Facebook showed its interest in blockchain space and announced its cryptocurrency called Libra. AWS announced the Amazon Blockchain Service on its cloud platform. In 2020, the price of Bitcoin increased to an all-time high of US$30,000. FinTech giant PayPal began experimenting with cryptocurrencies. As per the survey report submitted by Deloitte's Global Blockchain in 2020, around 40% of developers have implemented blockchain into production and 55% consider it as their next strategic initiative (Rani, Pareek, et al., 2023, February).

Web 2.0	Web 3.0
Read/Write Web	Portable Personal Web
Communities	Individuals
Sharing Content	Consolidating Dynamic Content
Blogs	Lifestream
AJAX	RDF
Wikipedia, google	Dbpedia, igoogle
Tagging	User engagement

FIGURE 6.6 Contrasting Web 2.0 applications with Web 3.0 decentralized applications.

In 2021, the cryptocurrency market experienced a tumultuous period for investors, with the value of Bitcoin reaching an all-time high of $64.829.14 before declining to a low of $35,000. The year 2022 began with an increase in the popularity of NFT and Metaverse. This let to increase in trading options and profit opportunities. A number of high-profile apparel brands such as Walmart, Adidas, Nike, Puma, Gucci, Ferrari, and Disney got involved in the metaverse. There are two main categories of Blockchain 4.0 applications: Web3 and the Metaverse. With Web3, people and crypto exchanges can reduce this risk and use Blockchain 4.0 to replace real money (Bhambri et al., 2023).

6.5.2 Web 3.0

The original internet was known as the "static web" because it was just a bunch of pages that did not have much to do with anything interactive. Web 1.0 was all about browsing static pages, and only a few people were in charge of generating content. Nowadays, we have the "interactive read-write" and social web, and lots of apps are designed so that anyone can become a creator and share their ideas with the world. For example, you can create videos on YouTube, photos on Flickr, photos on Instagram, and social media like Twitter.

A concise overview of the evolution of the web, from its inception in the Web 1.0 era to its current state in the Web 2.0 era, and what lies ahead in the future with the emergence of new technologies and the Web 3.0 era are presented by Goel et al. (2022). Figure 6.6 provides a comparison between the two, emphasizing parameters like computing, hosting, service layers, and storage. These alterations have led to a scenario where users have traded their privacy and control over data, rendering them susceptible to security breaches.

6.5.3 Metaverse

The metaverse is an emerging digital environment enhanced with 3D capabilities, leveraging technologies like virtual reality (VR), augmented reality (AR), and advanced internet and semiconductor technologies. The term "meta" implies "beyond," and "verse" signifies "universe," collectively forming the concept of the "metaverse." Also known as "virtual worlds," these metaverses serve as interactive platforms where individuals can engage, play, and interact with one another.

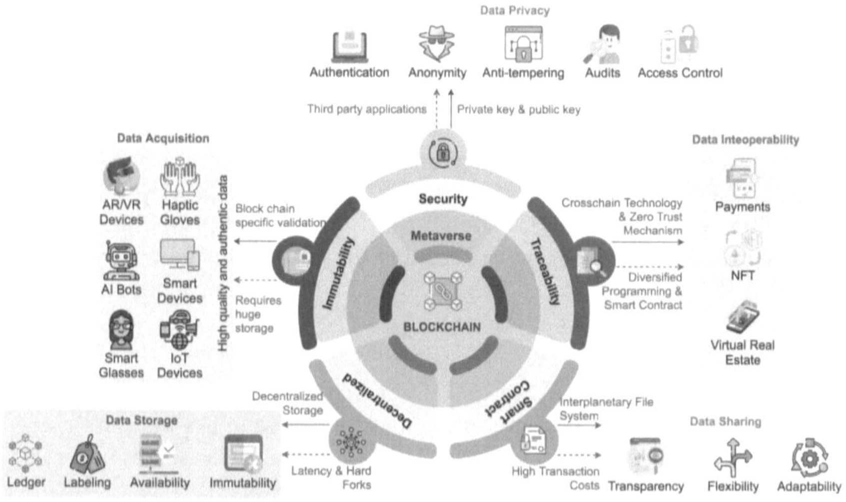

FIGURE 6.7 Blockchain for technical aspects in Metaverse.

Essentially, a metaverse offers a 3D virtual environment, providing people with unique experiences that transcend the limitations of the physical world.

The term "meta" is often used by computer industry to describe what many believe to be the next generation of the internet. It refers to a "one-size-fits-all" 3D virtual world that is shared, immersive, and persistent. Developers can build "metaverse" buildings, "metaverse" parks, "metaverse signs," and "metaverse objects." Creation of digital asset and smart contract is the main focus of Metaverse. Users can create and release their own unique digital assets using the metaverse Digital Asset System such as stocks, bonds, reward points, and more. The intersection of blockchain technology and metaverse development presents a comprehensive exploration of their symbiotic role (Figure 6.7).

6.6 TRENDS IN RESEARCH PUBLICATION ON BLOCKCHAIN TECHNOLOGY

A systematic mapping study was conducted by Yli-Huumo et al. (2016). The author collected many research papers on blockchain technology (Tanwar et al., 2022, September). A total of 41 primary papers were selected from scientific databases. Conducting a comprehensive analysis, Wang et al. (2021) scrutinized and examined a total of 2,451 papers published between 2013 and 2019. These papers were sourced from the Web of Science Core Collection database. There has been a notable transition from an emphasis on Bitcoin, cryptocurrencies, smart contracts, and the Internet of Things to a more focused analysis of distributed ledgers, challenges, and inefficiencies within the blockchain domain (Rani, Bhambri, et al., 2023).

6.7 EXPECTED BLOCKCHAIN PROSPECTUS IN 2023 AND FUTURE

The blockchain sector has revolutionized the economic and business landscape, largely because of its capability to provide privacy and security. There is an estimation from Business Research Company that the global market for blockchain services is expected to expand from an estimated US$3.28 billion per year in 2022 to US$4.7 billion per year in 2023, and to reach a total of US$19.76 billion per year by the end of 2027. The following are some of the key blockchain trends projected to emerge in 2023, which are expected to have a significant impact on the future of technology:

Decentralized Finance and Banking: DeFi stands for decentralized finance, which means you do not have to pay a fee or charge to use the bank's services. Instead, you hold your money in a safe digital wallet that is managed by smart contracts.

Online Payment Systems Using Cryptocurrencies: Blockchain-based payment systems leverage blockchain technology and cryptocurrencies to store transactional data in peer-to-peer networks, eliminating the necessity for centralized authorities like banks. This leads to a reduction in transaction costs, enhanced security, and a growing demand for high-performance systems capable of faster transaction times (Kataria et al., 2022).

Increased Business Adoption of Blockchain Technologies: The most significant trend in 2023 is the expansion of enterprise use cases for blockchains. The decentralized nature of blockchains provides enhanced security, transparency, and resilience to cyberattacks, which is one of the reasons more businesses are likely to adopt the technology.

Blockchain-Based Applications Are Expected to Grow Significantly in the Coming Years: In 2023, there will be high demand for blockchain software developers. There will be an increase in the demand for blockchain technology that can facilitate the development of robust applications to facilitate secure transactions, enhance Know Your Customer features, and more.

Blockchain is a game changer for the next tech revolution. It is not just for cryptocurrencies anymore. It is used in everything from healthcare to e-commerce to publishing to finance and insurance. Grand View Research predicts that the blockchain tech market will hit $1.432 billion by 2030, with a CAGR of 85.9%. With the Metaverse, NFTs, green initiatives, and other big blockchain trends in 2023, blockchain is here for the long haul.

In the book *The Fourth Industrial Revolution*, Prof. Klaus Schwab mentions the keyword blockchain as the future engineering technology of mankind. According to a forecast by research firm Gartner, by 2026, the business value added by blockchain will increase to over $360 billion. Then, by 2030, that will increase to more than $3.1 trillion.

6.8 CONCLUSION

The initiation of cryptocurrencies brought global recognition to blockchain technology, and its applications have expanded significantly. The progress from 1991 to 2023 is meticulously covered, encompassing cryptocurrencies, introduced mechanisms, and technological advancements. Blockchain 1.0 witnessed progressive development from 2008 to 2012 with the introduction of the Bitcoin cryptocurrency. Blockchain 2.0 evolved by introducing the concept of smart contracts with Ethereum and using Ether as a cryptocurrency. Blockchain 3.0 advanced with the development of decentralized applications, incorporating blockchain into various sectors such as healthcare, education, Internet of Things, and supply chain management. Blockchain 4.0 has gained momentum with the development of web 3.0 and the metaverse.

Looking ahead to 2023 and beyond, the prospects of blockchain technology are explored. The momentum is growing, with different industries actively experimenting with blockchain. The blockchain industry has unlocked numerous possibilities and opportunities at both economic and business levels. Based on current and future trends, there is a prediction that blockchain will lead a significant revolution in the coming decades.

REFERENCES

Agrawal, T. K., Angelis, J., Khilji, W. A., Kalaiarasan, R., & Wiktorsson, M. (2023). Demonstration of a blockchain-based framework using smart contracts for supply chain collaboration. *International Journal of Production Research, 61*(5), 1497–1516.

Alrubei, S., Rigelsford, J., Willis, C., & Ball, E. (2019, June). Ethereum blockchain for securing the Internet of Things: Practical implementation and performance evaluation. In *2019 International Conference on Cyber Security and Protection of Digital Services (Cyber Security)* (pp. 1–5). UK: IEEE.

Armknecht, F., Karame, G. O., Mandal, A., Youssef, F., & Zenner, E. (2015). Ripple: Overview and outlook. In *Trust and Trustworthy Computing: 8th International Conference, TRUST 2015, Heraklion, Greece, August 24–26, 2015, Proceedings 8* (pp. 163–180). New York, NY: Springer International Publishing.

Awan, S. H., Ahmed, S., Nawaz, A., Sulaiman, S., Zaman, K., Ali, M. Y., . . . Imran, S. (2020). BlockChain with IoT, an emergent routing scheme for smart agriculture. *International Journal of Advanced Computer Science and Applications, 11*(4), 420–429.

Bali, V., Bali, S., Gaur, D., Rani, S., & Kumar, R. (2023). Commercial-off-the shelf vendor selection: A multi-criteria decision-making approach using intuitionistic fuzzy sets and TOPSIS. *Operational Research in Engineering Sciences: Theory and Applications, 6*(2).

Benedict, S., Rumaise, P., & Kaur, J. (2019, December). IoT blockchain solution for air quality monitoring in SmartCities. In *2019 IEEE International Conference on Advanced Networks and Telecommunications Systems (ANTS)* (pp. 1–6). Goa: IEEE.

Bhambri, P., & Rani, S. (2024). Challenges, opportunities, and the future of industrial engineering with IoT and AI. *Integration of AI-Based Manufacturing and Industrial Engineering Systems with the Internet of Things*, 1–18.

Bhambri, P., Rani, S., Balas, V. E., & Elngar, A. A. (Eds.). (2023). *Integration of AI-Based Manufacturing and Industrial Engineering Systems with the Internet of Things*. USA: CRC Press.

Cai, W., Wang, Z., Ernst, J. B., Hong, Z., Feng, C., & Leung, V. C. (2018). Decentralized applications: The blockchain-empowered software system. *IEEE Access, 6*, 53019–53033.

Chauhan, M., & Rani, S. (2021). Covid-19: A revolution in the field of education in India. *Learning How to Learn Using Multimedia*, 23–42.

Goel, A. K., Bakshi, R., & Agrawal, K. K. (2022). Web 3.0 and decentralized applications. *Materials Proceedings*, *10*, 8.

Honar Pajooh, H., Rashid, M., Alam, F., & Demidenko, S. (2021). Hyperledger fabric blockchain for securing the edge internet of things. *Sensors*, *21*(2), 359.

Huang, Y., Wang, B., & Wang, Y. (2021). Research and application of smart contract based on ethereum blockchain. In *Journal of Physics: Conference Series* (Vol. 1748, No. 4, p. 042016). UK, England: IOP Publishing.

Idrees, S. M., Nowostawski, M., & Jameel, R. (2021). Blockchain-based digital contact tracing apps for COVID-19 pandemic management: Issues, challenges, solutions, and future directions. *JMIR Medical Informatics*, *9*(2), e25245.

Kataria, A., Agrawal, D., Rani, S., Karar, V., & Chauhan, M. (2022). Prediction of blood screening parameters for preliminary analysis using neural networks. In *Predictive Modeling in Biomedical Data Mining and Analysis* (pp. 157–169). USA: Academic Press, Elsevier.

Kataria, A., Puri, V., Pareek, P. K., & Rani, S. (2023, July). Human activity classification using G-XGB. In *2023 International Conference on Data Science and Network Security (ICDSNS)* (pp. 1–5). Tiptur: IEEE.

Kumar, P., Banerjee, K., Singhal, N., Kumar, A., Rani, S., Kumar, R., & Lavinia, C. A. (2022). Verifiable, secure mobile agent migration in healthcare systems using a polynomial-based threshold secret sharing scheme with a blowfish algorithm. *Sensors*, *22*(22), 8620.

Li, D., Wong, W. E., & Guo, J. (2020, January). A survey on blockchain for enterprise using hyperledger fabric and composer. In *2019 6th International Conference on Dependable Systems and Their Applications (DSA)* (pp. 71–80). China: IEEE.

Mayer, A. H., da Costa, C. A., & Righi, R. D. R. (2020). Electronic health records in a blockchain: A systematic review. *Health Informatics Journal*, *26*(2), 1273–1288.

Ncube, T., Dlodlo, N., & Terzoli, A. (2020, November). Private blockchain networks: A solution for data privacy. In *2020 2nd International Multidisciplinary Information Technology and Engineering Conference (IMITEC)* (pp. 1–8). South Africa: IEEE.

Ocheja, P., Flanagan, B., Ueda, H., & Ogata, H. (2019). Managing lifelong learning records through blockchain. *Research and Practice in Technology Enhanced Learning*, *14*(1), 1–19.

Pongnumkul, S., Siripanpornchana, C., & Thajchayapong, S. (2017, July). Performance analysis of private blockchain platforms in varying workloads. In *2017 26th International Conference on Computer Communication and Networks (ICCCN)* (pp. 1–6). Canada: IEEE.

Puri, V., Kataria, A., Solanki, V. K., & Rani, S. (2022, December). AI-based botnet attack classification and detection in IoT devices. In *2022 IEEE International Conference on Machine Learning and Applied Network Technologies (ICMLANT)* (pp. 1–5). Salvador: IEEE.

Rani, S., Bhambri, P., & Kataria, A. (2023). Integration of IoT, big data, and cloud computing technologies. In *Big Data, Cloud Computing and IoT: Tools and Applications*. USA: CRC Press.

Rani, S., Kataria, A., & Chauhan, M. (2022). Cyber security techniques, architectures, and design. In *Holistic Approach to Quantum Cryptography in Cyber Security* (pp. 41–66). USA: CRC Press.

Rani, S., Kaur, J., & Bhambri, P. (2023). Technology and gender violence: Victimization model, consequences and measures. In *Communication Technology and Gender Violence* (pp. 1–19). Cham: Springer International Publishing.

Rani, S., Kumar, S., Kataria, A., & Min, H. (2023). SmartHealth: An intelligent framework to secure IoMT service applications using machine learning. *ICT Express*, *10*(2), 420–425.

Rani, S., Mishra, A. K., Kataria, A., Mallik, S., & Qin, H. (2023). Machine learning-based optimal crop selection system in smart agriculture. *Scientific Reports*, *13*(1), 15997.

Rani, S., Pareek, P. K., Kaur, J., Chauhan, M., & Bhambri, P. (2023, February). Quantum machine learning in healthcare: Developments and challenges. In *2023 IEEE International Conference on Integrated Circuits and Communication Systems (ICICACS)* (pp. 1–7). Raichur: IEEE.

Singh, H., & Rani, S. (2023). Flex sensor integrated smart strap to verify correct wearing of the face mask. *IEEE Sensors Journal*, *24*(2), 2020–2027.

Sudevan, S., Barwani, B., Al Maani, E., Rani, S., & Sivaraman, A. K. (2021). Impact of blended learning during Covid-19 in Sultanate of Oman. *Annals of the Romanian Society for Cell Biology*, 14978–14987.

Tanwar, R., Chhabra, Y., Rattan, P., & Rani, S. (2022, September). Blockchain in IoT networks for precision agriculture. In *International Conference on Innovative Computing and Communications: Proceedings of ICICC 2022* (Vol. 2, pp. 137–147). Singapore: Springer Nature Singapore.

Turkanović, M., Hölbl, M., Košič, K., Heričko, M., & Kamišalić, A. (2018). EduCTX: A blockchain-based higher education credit platform. *IEEE Access*, *6*, 5112–5127.

Vaneeta, M., Deepa, S. R., Sangeetha, V., Naganna, K., Vasisht, K. S., Ashwini, J., . . . Srividya, H. R. (2023). Tamper proof air quality management system using blockchain. *International Journal of Advanced Computer Science and Applications*, *14*(2).

Wang, G., Zhang, S., Yu, T., & Ning, Y. (2021). A systematic overview of blockchain research. *Journal of Systems Science and Information*, *9*(3), 205–238.

Westerlund, M., Nene, S., Leminen, S., & Rajahonka, M. (2021). An exploration of blockchain-based traceability in food supply chains: On the benefits of distributed digital records from farm to fork. *Technology innovation management review*, *12*(3).

Yli-Huumo, J., Ko, D., Choi, S., Park, S., & Smolander, K. (2016). Where is current research on blockchain technology? A systematic review. *PLoS One*, *11*(10), e0163477.

7 Developing Smart Cities by Integrating Blockchain-Based GRNN with CSO-Transformed Paillier Encryption Model

Vasantha M., Sudarsanan D., Santosh M., and Madhura G.K.

7.1 INTRODUCTION

Sustainable transportation, energy, greenhouse gas emissions, health, and quality of life are just some of the social concerns that smart cities are tackling [1]. New public goods are made available to citizens as a result of the pervasiveness of IT-based smart solutions. Therefore, it has become a new information source for city administrations for smart open government, on the basis of which municipal development proposals and strategic choices are formed [2]. A smart open government strategy seeks to include citizens using novel means, delivering services that citizens can trust to be open and honest. [3]. Citizen engagement or participation, for example, the smarticipate initiative [4], refers to the involvement of citizens in city sessions, opinions, surveys, and call-for-comments on online projects. Civic officials may then co-create and modify civic infrastructure and services to better meet the needs and requirements of the local population.

Current citizen engagement solutions are centralized and tailored to address narrow problem spaces [5]. However, transferring and processing citizen involvement data at a data center or cloud might have a significant carbon footprint across functioning urban regions (i.e., districts). Open public services and the data they generate are also susceptible to security and privacy breaches [6]. Therefore, it is crucial that IT infrastructure in smart cities be redesigned and configured so that it can efficiently handle data storage; for example, by protecting citizen participation data at the district/unit level and then relaying the processed data to city administrators to use in city planning [7]. The information technology infrastructure of a smart city should protect the confidentiality of all data and services it manages, even if they are

 DOI: 10.1201/9781003460367-7

provided outside of the city's borders. Delegation of authority from city hall to local councils or administrative districts is a common practice [8]. This implies that in a district, the town councils or local district administration are responsible for managing some services, while the main City Administration is responsible for providing and making decisions on significant or vital services [9]. The City Administration manages the refined and resulting data, while the local data is processed locally to protect user privacy. We accounted for this restriction in our approach by allocating public and government access based on location.

Researchers in industry have paid a lot of attention to the privacy concerns that arise while training ML classifiers. To ensure the confidentiality of user information, current solutions [10] use cryptographic methods or differential privacy. They often ignore issues about data quality and ownership [11] and operate under the premise that training data can be gathered safely from many data providers for subsequent analysis. However, prospective assaults mean this is not always the case in practice. In this work, we use blockchain technology to create a safe environment for exchanging information, bridging the gap between ideal assumptions and practical realities. Blockchain is simply a decentralized file system that facilitates the sharing of immutable documents across many entities [12]. Data auditing, made possible by blockchain's permanent and immutable records, can verify the rightful owners of stored data. This makes it easier to determine the precise value added by each data contributor and to implement incentive schemes to promote the widespread dissemination of training data [13].

Integrating blockchain into the DL training process is a promising but yet difficult undertaking. The first difficulty is coming up with a training data format that is compatible with blockchain and secure enough to guarantee the anonymity of each individual provider [14]. The second difficulty is creating a training algorithm that can use blockchain data to produce private and accurate GRNN classifiers. The research uses blockchain methods to enable safe and trustworthy IoT data exchange, therefore resolving the aforementioned problems. By using its own private key, each IoT data source may encrypt data instances locally before recording them on the blockchain in a specifically designed transaction. The key objective is to supply a groundbreaking CSO-based coupled transformed Paillier and KLEIN algorithm where the model builds a trustworthy GRNN training algorithm with just two interactions in each iteration and no need for a third-party guarantor of security. Extensive experiments are performed to demonstrate that our technique may be used to train GRNN classifiers safely with a reasonable degree of accuracy.

Here is how the rest of the chapter is laid out: A brief overview of the background literature is provided in Section 7.2. An explanation of the proposed model is provided in Section 7.3. An analysis of experiments is provided in Section 7.4. Section 7.5 provides a summary, Section 7.6 presents performance validation of the model, and Section 7.7 concludes the chapter.

7.2 RELATED WORKS

An efficient decentralized healthcare financial system that protects users' privacy and is optimized for mobile devices has been proposed by Singh et al. [15]. The suggested architecture prioritizes a non-interactive zero-knowledge proof, which drastically lowers the cost of inter-device communication. In this chapter, we provide a micro-level overview of the system framework and its application to a healthcare finance system. However, it is easily generalizable to broader monetary systems. Our system framework is lightweight and fast, validating transactions in milliseconds with the use of more efficient zero-knowledge-based proofs. In addition to being auditable, the suggested healthcare finance system for portable computing devices does not disclose any unnecessary details.

Using recent developments in AI methods, Kumar et al. [16] established a privacy preservation model for IIoT. IIoT data sanitization and restoration are the two primary phases of the proposed solution. In order to stop the leaking of private data, IIoT uses a process called data sanitization. In addition, a novel Grasshopper method is used in the planned sanitization phase to generate keys that are both efficient and secure. Optimal keys were generated using a multi-objective function that took into account characteristics such as degree of change, concealment rate, correlation coefficient between original and restored data, and information preservation rate. In terms of a wide range of performance measures, the simulation result demonstrates that the suggested model is superior to the state-of-the-art models. When compared to JA, GWO, GOA, and BHO, the suggested G-BHO algorithm achieves a 1%, 15%, 2%, and 12.6% better result, respectively, in terms of privacy protection.

Background of IoT systems and privacy and security measures are described by Safaei Yaraziz et al. [17], who also advocate privacy models for various layers of IoT applications and discuss (a) ways to protecting privacy in IoT-based schemes and (b) current privacy solutions. Our research shows that innovative approaches to privacy protection—including blockchain technology, artificial intelligence, data minimization, and encryption—can have a significant effect on securing personal information. In addition, it makes logical that consumers may better safeguard their privacy if smart gadgets acquire, keep, and share less data about them. Because of the importance of privacy in these networks, this research presents a data minimization strategy based on machine learning [18–24].

For the sake of confidentiality, Sharma et al. [25] have offered in this work a distributed application (DA) that would use blockchain technology to issue and store medical certifications. Medical records may be created and issued by healthcare facilities, verifiers, and regular authorities with the use of a distributed application that acts as a bridge between the blockchain network and these system objects. In addition, it provides safety by allowing users to set restrictions using a variety of smart contracts. Using the Etherscan instrument, we undertake a number of experimental experiments to calculate the overhead, latency, and processing time of the proposed method. Here, we evaluate the proposed system's latency, throughput, and reaction time efficiency to those of the existing systems. The experimental findings and analysis of existing methods demonstrate that the suggested work is superior [26–29].

The differential privacy mechanism provided by Gao et al. [30] build a formal economic model for IoT data mining and a privacy-preserving data mining protocol for the Internet of Things (IoT-PPDM). To model the multilateral interaction process inherent in IoT data mining, we turned to non-cooperative game theory. A dynamic incentive system is developed to strike a compromise between privacy protection and data mining needs in order to increase involvement from all relevant stakeholders. Furthermore, we theoretically verify the viability of IoT-PPDM and explain the value of all involved parties. IoT-PPDM with a dynamic incentive mechanism has been shown experimentally to maximize advantages for all parties involved while also discouraging any unreasonable actions on the part of any of them [31–34].

Using Homomorphic Paillier Encryption, Gheisari et al. [35] present a privacy-preserving data aggregation architecture called PPDMIT. By running extensive simulations, we were able to determine that the suggested privacy-preserving architecture improves upon the state of the art in terms of both data aggregation efficiency and privacy protection. In addition, we determined that our suggested architecture effectively safeguards sensitive information in IoT settings. In particular, compared to LPDA, our solution was 8.096% better, and compared to PPIOT, it was 6.508% better.

7.3 PROPOSED SYSTEM

Here, we lay out the system architecture, threat model, and design goals for the challenge of safe GRNN training over encrypted data acquired from multiple sources [36, 37].

7.3.1 System Model

We imagine an Internet of Things (IoT) ecosystem, including components such as IoT gadgets, IoT data suppliers, an IoT platform powered by blockchain technology, and IoT data analyzers.

- Wireless or wired networks like ZigBee, 3G/4G, and Wi-Fi enable IoT devices to sense and communicate vital data. From weather reports to biometric data, the collected information may be put to use in a variety of ways in today's smart cities. Due to their limited processing capabilities, IoT devices will not contribute to the data exchange and analysis procedures.
- All data from IoT devices is collected by the respective IoT data providers. Sensitive information is often contained in IoT data since it is a significant asset to data providers. So, each data source uses partly homomorphic encryption to secure its IoT data before adding it to the distributed ledger.
- To record the encrypted IoT data collected from all data sources, the blockchain-based IoT platform acts as a distributed database. We can guarantee the integrity and veracity of shared IoT data using the built-in consensus process.
- The goal of IoT data analysts is to have a thorough understanding of the IoT data stored on the blockchain-based platform by making use of cutting-edge analytic tools. In order to collect the parameters of training GRNN classifiers, data analysts need to interact with the necessary data providers.

7.3.2 Threat Model

The system model suggests that there are many different kinds of dangers to each entity type and their interactions. As our goal is to build a privacy-preserving strategy for training GRNN models across various IoT providers, we focus solely on the risks to data privacy that arise from the exchange of information between providers and analysts.

The data analyst is viewed as a fair and inquisitive foe, meaning it would follow the predesigned DL training protocols in good faith, but it would also be interested about the substance of the data and try to learn more by decrypting and analyzing the intermediate data of computation.

Based on the type of sensitive information that may be collected by the data analyst, we take into account two threat models with varying attack capabilities that are widely used in the literature:

- *Model of Known Ciphertext:* The IoT data analyst has no other way of accessing the data but through the blockchain. The IoT data analyst may also keep track of the iteration step and descent gradient that are produced by running the secure training algorithm.
- *Model of Contextual Knowledge Known to Exist:* The IoT data analyst in this more robust model is believed to have more information at their disposal than is available in the known ciphertext model.

IoT data analysts, in particular, can deduce the private information of other IoT data providers by conspiring with other providers.

7.3.3 Training the GRNN Model

The year 2014 saw the introduction of a new type of RNN called Gated Recurrent Unit (GRU) networks. Like the LSTM, the GRU can govern the flow of info without the need of a dedicated memory cell, and its construction is simpler. Voice recognition tasks, including voice augmentation, speech activity detection, and text recommendation [38], have benefited from the use of GRUs. To keep tabs on the status of sequences, a GRU has two gates—the update gate and the reset gate. The update gate determines what data is kept and what is overwritten, balancing the two. The reset gate, on the other hand, regulates the weight of history in the present. The GRU stands out from the LSTM in that it does not rely on internal memory cells to record dependencies of lengthy time series. This helps speed up the process of resolving the vanishing gradient problem.

In LSTM, the update gate performs a function analogous to that of the forget gate and the input gate. Both the fresh input and the concealed state from the preceding timestep are multiplied by their respective weights. The function is used to map the combined contributions of the two candidates from zero to one.

$$Z_t = o\left(X_t W^z + h_{t-1} U^z + b_z\right) \tag{7.1}$$

where Z_t is an illustration of the update gate, X_t is the time step t input vector, and $h_{(t-1)}$ is the time step $t-1$ output. The W^z represents the input layer weight, whereas the U^z represents the recurrent weight. The input layer bias is represented by b_z. The extent to which the past should be forgotten is set by the reset gate. The equation for it is quite close to the update gates. Multiplying the novel input unit by its associated weight W^r, U^r, and then passing the product via a sigmoid function, produces the hidden state. This is the result of the reset gate's output:

$$R_t = o\left(X_t W^r + S_{t-1} U^r + b_r\right) \tag{7.2}$$

The novel memory contact to store the info of the previous state. The input gate multiplied by its weight Hadamard (⊙) should be gate (R_t) and previous out (h_{t-1}). This will let the network to relevant past info. Current is intended as shadow.

$$\tilde{h}_t = tanh\left(X_t W + U\left(R_t \odot h_{t-1}\right)\right) \tag{7.3}$$

$$h_t = Z_t \odot h_{t-1} + \left(1 - Z_t\right) \odot o\left(\tilde{h}_t\right) + b_h \tag{7.4}$$

The update gate evaluates the state of memory and prior operations to determine what data should be gathered. The new hidden state is produced at the current time step and goes via an activation function called sigmoid (o).

7.3.4 Blockchain Systems

Blockchain is a decentralized, public ledger in the form of a linked list of blocks first developed to record Bitcoin and other cryptocurrency transactions. It facilitates trustworthy exchanges between parties where no trust exists. Hyperledger, Ethereum, and EOS are just some of the blockchain platforms that have been developed and used in various contexts throughout the past year. Blockchain platforms may be loosely divided into three groups, public blockchains, private blockchains, and according to the access limitation on blockchain users.

Blockchain's many useful properties make it an ideal platform for trustworthy information exchange:

Totally Decentralized: Blockchain is a decentralized ledger that operates on a peer-to-peer network without a centralized authority or trusted third party. It is impossible to lose any information stored in the ledger since several copies of the data are kept in the system.

Safe Against Tampering: To regulate who gets to add new blocks to the ledger, blockchain relies on consensus techniques like Proof of Work (PoW). Because of the prohibitive computational cost of modifying data, it is frequently impossible to modify information stored in blocks.

Capability of Being Tracked: In a blockchain system, all participants may readily verify any and all transactions between any two parties. In real time, the data owner can see how their data is being utilized by others and how much they are being paid for each usage.

Let $N > 1$ be a number. Then Z_N^* is an abelian group below multiplication modulo N. Describe $\varphi(N)\text{def}\left[Z_N^*\right]$, the instruction of the group Z_N^*.

While blockchain has a lot going for it, it is far from perfect when it comes to functioning as a platform for exchanging data because of the privacy issues that might arise from using it. In the original implementation, all transactions are stored in blocks as plaintexts, leaving all participants, including attackers, with full access to potentially sensitive information in transactions [19]. Therefore, while using blockchain as a data exchange platform, it is important to properly handle security and privacy concerns.

7.3.5 Design Goals

To steal the privacy of other participants, we allow any two or more IoT data suppliers to collude with IoT data analyzer. Here are the presumptions we are making. Each player takes on the role of a trustworthy opponent who follows the rules of the protocol but who also could be intrigued about the sensitive data of other domains. It is possible for two or more participants to work together. They are passive adversaries that do their best to infer private information about others based on the values they discover while still adhering to the protocol. Our method is designed to keep personal information safe while safely training a GRNN model.

7.3.6 Transformed Paillier Encryption

This system may be used by many people, however each user (or data owner) would like to encrypt their image with a different pair of keys. These keys are created for each user by the key management authority following an initiation procedure. The first layer of encryption is created using the original key pair. Paillier encrypts the second key pair with KLEIN. The key management authority supplies the encryption and decryption keys. To symbolize the launch, the initialization procedure is implemented. Pick two large prime numbers, pr_1 and pr_2, at random and separately, such that $gc\,d\left(pr_1 * pr_2\left(pr_1 - 1\right)\left(pr_2 - 1\right)\right)1$.

$$\begin{aligned} Sk &= pr_1 * pr_w, \\ \Lambda &= lcm\left(pr_1 - 1, pr_2 - 1\right) \\ M &= \left(L\left(g^{\Lambda} \bmod Sk^2\right)\right)^{-1} \bmod Sk \end{aligned} \tag{7.5}$$

Here, $g \in Z$, the public keys are Sk and g, and the private keys for decryption are Λ and M. Let the image pixel pr_1 be encrypted, where Λ is the demanded leakage function, $0 \le pr_1 \le Sk$, and an random integer rand integers 0 through n, where n is the desired number. Using these two numbers, the encryption process will be represented by the number (6). The encrypted picture pixel is denoted by E, the minimum polynomial by min p, and the random number n by the notation rand.

$$E = \min p \cdot \text{rand}^n \bmod Sk^2 \tag{7.6}$$

This decryption operation is represented by the secret key (7), which the user will use to accomplish the second stage of encryption when the encryption process is complete.

$$D = L\left(c^{\wedge} \bmod Sk^{2}\right).\, M \bmod \tag{7.7}$$

where D is the image's decoded pixel and c is the original, plaintext.

7.3.7 KLEIN Encryption Process

KLEIN employs a standard substitution–permutation network (SPN) architecture, similar to that of other advanced block ciphers like AES and PRESENT. The predicted NR rounds for KLEIN64/80/96 are 12/16/20, providing an adequate security cushion and asymmetric iteration. All practical block ciphers make use of separate key schedules to translate relatively tiny master keys into subsequent transformations. KLEIN is used to construct block authentication codes even when keys are often modified, requiring a flexible key schedule. On the other hand, key scheduling takes into account an appropriate amount of security complexity. The following explains how KLEIN's key scheduling works to mitigate any negative effects of related key deficiencies on performance:

(1) *Input:* A 64/80/96-bit for KLEIN-64/80/96.

(2) *Key Scheduling:* When $i = 1$, the initial subkey $sk1 = mk = sk_0^1 \| sk_1^1 \| \ldots \| sk_t^1$, where $t = 7/9/11$ for KLEIN-64/80/96. For KLEIN-64, the $(i + 1)$th subkey sk^{i+1} can result from the ith subkey sk^i as follows:

   (i) Divide the ith subkey sk^i into two tuples, such that $t_1 = \left(sk_0^i, sk_1^i, \ldots, sk_{\lfloor t/2 \rfloor}^i\right)$ and $t_2 = \left(sk_{[t/2]}^i, sk_{\left\lceil \frac{t}{2} \right\rceil + 1}^i, \ldots, sk_t^i\right)$ for the next step.

   (ii) Cycling left shift one byte site in (t_1, t_2), obtain $t_2' = \left(sk_{\left\lceil \frac{t}{2} \right\rceil + 1}^i, \ldots, sk_{[t/2]}^i, sk_0^i\right)$ and $t_2' = \left(sk_{\lfloor t/2 \rfloor + 1}^i, \ldots, sk_t^i, sk_{\lfloor t/2 \rfloor}^i\right)$ for the next step.

   (iii) Swap the tuple (t_1, t_2) with a Feistel-like structure, such that $t_1'' = t_2'$ becomes the left tuple, while $t_2'' = t_1' \oplus t_2^1$ develops the right tuple.

   (iv) XOR round pawn i with the third byte in the left tuple t_1'', and we extract the second and the third bytes of the right tuple t_2'' by using the KLEIN S-box.

(3) To get the desired result, repeat the preceding procedure for a variety of key lengths and discard the first 64 bits of subkey sk^i before doing the ith cycle of transformation.

KLEIN's key scheduling may work with a wide variety of key sizes. As a memory-saving measure, KLEIN subkeys can be produced via iterative transformations. When optimizing the sensor's performance, KLEIN is on the fly than traditional optimizations, which require computing all subkeys in advance. The primary space optimization performed by CSO was sensitive to the amount of iteration rounds.

7.3.8 Cuckoo Search Optimization

Cuckoo search optimization (CSO) [39] is a proposed metaheuristic procedure inspired by the Lévy flying behavior of many species of fruit flies and birds, as well as the important brood parasite behaviors of some cuckoo species. Some species of cuckoo are known to aggressively reproduce by birds. It is possible that some other species of bird is the host. If it discovers that the eggs are its own, it may either remove the foreign eggs or abandon the nest altogether in favor of building a new one. The CS method and its variants have been shown to be effective in solving a wide variety of practical optimization problems.

Here are three CS regulations cited in [39]:

1. A cuckoo only lays one egg at a time, and she will place it in whatever nest she may find.
2. The best nest and eggs will be passed down from generation to generation.
3. The chance that a host bird will find the cuckoo egg is pa 2 [0, 1] when there are a certain number of nests available. The laying bird might either discard the egg or abandon the nest altogether to start over.

Each egg is a potential strategy that has been laid in a nesting for a certain problem and must be minimized, much like the cuckoo's song. It has been hypothesized that an egg is of higher quality and has a higher chance of survival or development if it is an exact replica of the eggs laid by the host bird. The cuckoo-like flying pattern of a Lévy flight is utilized to simulate the bird's hunt for a comfortable place to nest. If the newly laid cuckoo egg is of a greater quality, it will substitute the original eggs in the nests (or the old solution). The worst nests are thrown out (pa), and a new one is built at random.

In the CSO method, each cuckoo is tweaked with the help of Lévy flights, a modernized type of random walking.

$$x_i^{t+1} = x_i^t + s \tag{7.8}$$

where x_i^{t+1}, S is the ith cuckoo at repetition $t + 1$, and s is the step scope of the Lévy flights. It may be strong-minded using Mantegna's technique as shadows:

$$s = \alpha\left(x_i^t - x_{\text{best}}^t\right) \oplus \text{Lévy}(\beta) \tag{7.9}$$

$$\text{Lévy}(\beta) = \frac{u}{|v|^{1/\beta}} \tag{7.10}$$

where α is a continuous and β is the Lévy advocate, where $0 < \beta \leq 2$. The operator $\oplus$ is the elemental end outcome. x_i^t and x_{best}^t are the ith cuckoo and the finest cuckoo of group t, respectively. $\text{Lévy}(\beta)$ distribution, u and v, are center, $u \sim N(0,\sigma_u^2)$ and $v \sim N(0,\sigma_v^2)$. Their standard deviations are distinct as shadows:

$$\sigma_u = \left[\frac{\Gamma(1+\beta)\cdot \sin\left(\frac{\pi\beta}{2}\right)}{\Gamma\left(\frac{1+\beta}{2}\right)\cdot \beta.2^{(\beta-1)/2}} \right]^{1/\beta}, \sigma_v = 1 \tag{7.11}$$

where Γ is the gamma. Procedure 1 presents the description of CSO.

Algorithm 1. Cuckoo Search via Lévy Flights

1: set up N host nests, $N = \{x_1, x_2, x_3 \ldots, x_N\}$

2: while (t < T_{max}) or (stop criterion) do

3: *Choose a cuckoo at random from Lévy flights and assess its fitness* F_i

4: *Select one nest (say j) at random among N nests.*

5: *if* $F_j < F_i$ *then*

6: *Substitute the new solution for j.*

7: *end if*

8: A portion p_a of novel nests are built after the worst ones have detached.

9: *keep onto the best solutions*

10: *Determine the current best by ranking the solutions.*

11: *end while*

7.4 RESULTS AND DISCUSSION

The performance of the suggested technique is measured against that of other alternatives using symmetric cryptography. The lab uses a Xen Server-based virtual server environment hosted on the cloud. The cloud server has a Core I7 CPU running at 4.8 GHz, while the client PC has a Core I5 with 8 GB of RAM.

7.5 VALIDATION ANALYSIS OF PROPOSED MODEL

Calculate the encrypting time for the various input sizes (10, 20, 30, 50, and 100 kB) listed in Table 7.1 and Figure 7.1.

Despite the linear increase in encryption time with file size and character count, the results suggest that decryption time is less. This demonstrates that the proposed method requires significantly less computing power. The proposed method is faster than earlier genetic algorithms in encrypting plain text. Decryption time is an indication of how long it takes to recover the data after encryption. Table 7.1 shows that the

TABLE 7.1
Processing Period of the Proposed Model

Plaintext Size (kB)	Encryption Time (s)	Decryption Time (s)
5	0.57	0.371
10	1.08	0.712
20	1.92	1.581
30	2.69	1.504
40	3.59	2.49
Average	1.874	1.471

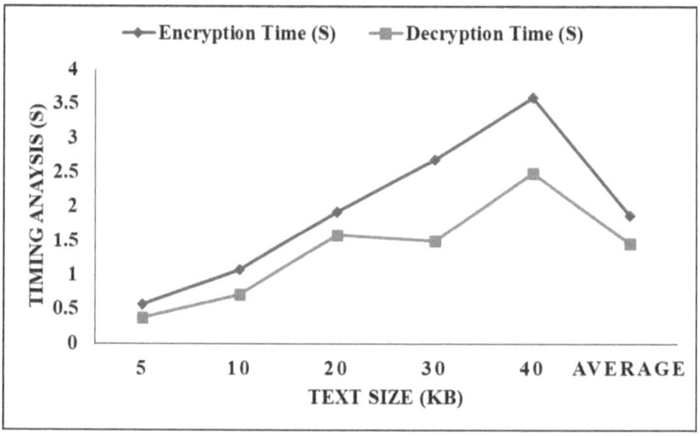

FIGURE 7.1 Graphical comparison of proposed encryption model.

proposed method requires less time to decode than the previous encryption methods. In conclusion, the proposed paradigm is useful for establishing encrypted communication and is simple to apply.

7.6 PERFORMANCE VALIDATION ON PROPOSED CLASSIFIER

Table 7.2 provides the comparative analysis of various classifiers in terms of different metrics.

Table 7.2 presents the experimental analysis of the proposed model. The MLP model attained the accuracy score of 87.72 and the precision rate of 88.14 and the recall range as 87.92 and the F-measure range as 88.67, correspondingly (Figure 7.2). The AE model attained the accuracy score as 89.17 and the precision rate as 90.91 and the recall range as 89.69 and the F-measure range as 89.33, correspondingly. Then the DBN model attained the accuracy score as 90.28 and the precision rate as 91.17 and the recall range as 91.66 and the F-measure range as 92.24, correspondingly. Then the CNN model attained the accuracy score as 91.78 and the precision

TABLE 7.2
Experimental Analysis of Proposed Model

Classification	Accuracy (%)	Precision (%)	Recall (%)	F-measure (%)
MLP	87.72	88.14	87.92	88.67
AE	89.17	90.91	89.69	89.33
DBN	90.28	91.17	91.66	92.24
CNN	91.78	93.94	94.61	93.86
RNN	92.34	94.45	96.78	94.36
GRNN	**95.97**	**96.84**	**98.24**	**95.13**

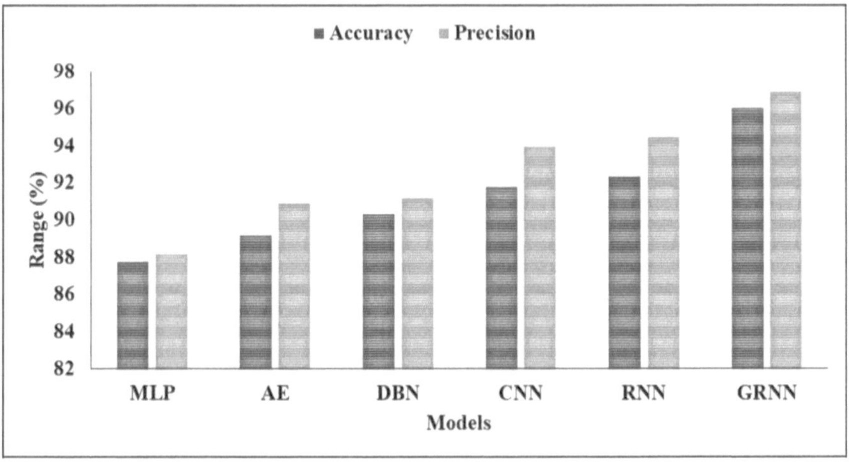

FIGURE 7.2 Graphical investigation of proposed classifier.

rate as 93.94 and the recall range as 94.61 and the F-measure range as 93.86, correspondingly. Then the RNN model attained the accuracy score as 92.34 and the precision rate as 94.45 and the recall range as 96.78 and the F-measure range as 94.36, correspondingly. Then the GRNN model attained the accuracy score as 95.97 and the precision rate as 96.84 and the recall range as 98.24 and the F-measure range as 95.13, correspondingly (Figure 7.3).

7.7 CONCLUSIONS

To address the issues of data integrity, we present secure GRNN, a novel privacy-preserving GRNN training arrangement that uses blockchain practices to construct a secure GRNN training procedure for use in multi-part situations in which IoT data is composed from multiple data breadwinners. An effective and accurate privacy-protecting GRNN training method is built using a transformation of Paillier and

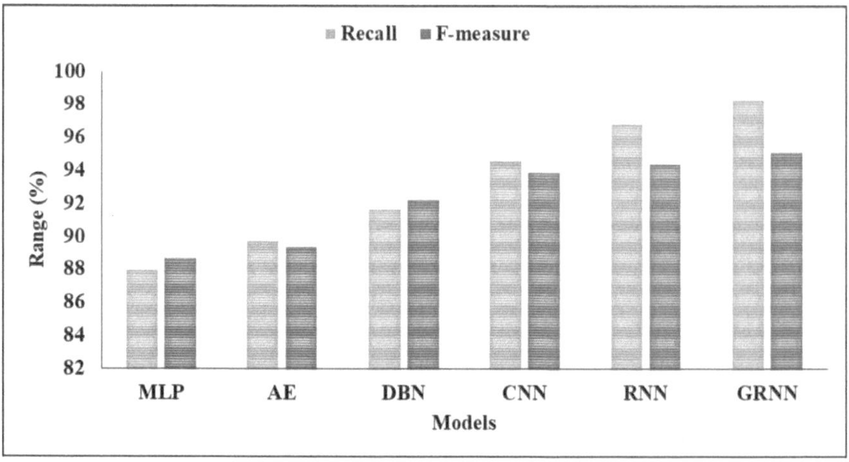

FIGURE 7.3 Validation investigation of various ML/DL techniques.

KLEIN. The CSO algorithm chooses the best key space for the KELIN. We showed that secure GRNN works and is safe to use. Our long-term goal is to provide a unified platform for building multi-part encrypted datasets suitable for deep learning privacy-preserving training techniques.

REFERENCES

1. Peng, L., Feng, W., Yan, Z., Li, Y., Zhou, X., & Shimizu, S. (2021). Privacy preservation in permissionless blockchain: A survey. *Digital Communications and Networks*, 7(3), 295–307.
2. Miyachi, K., & Mackey, T. K. (2021). hOCBS: A privacy-preserving blockchain framework for healthcare data leveraging an on-chain and off-chain system design. *Information Processing & Management*, 58(3), 102535.
3. Iftikhar, Z., Javed, Y., Zaidi, S. Y. A., Shah, M. A., Iqbal Khan, Z., Mussadiq, S., & Abbasi, K. (2021). Privacy preservation in resource-constrained IoT devices using blockchain—A survey. *Electronics*, 10(14), 1732.
4. Qiao, Y., Lan, Q., Zhou, Z., & Ma, C. (2022). Privacy-preserving credit evaluation system based on blockchain. *Expert Systems with Applications*, 188, 115989.
5. Li, T., Wang, H., He, D., & Yu, J. (2022). Blockchain-based privacy-preserving and rewarding private data sharing for IoT. *IEEE Internet of Things Journal*, 9(16), 15138–15149.
6. Yang, Y., Chen, Q., Huang, T. Y., & Pareek, P. K. (2023). Application research of K-means algorithm based on big data background. In: *2023 IEEE International Conference on Integrated Circuits and Communication Systems (ICICACS)* (pp. 1–5). Raichur: IEEE. https://doi.org/10.1109/ICICACS57338.2023.10099551
7. Hasan, O., Brunie, L., & Bertino, E. (2022). Privacy-preserving reputation systems based on blockchain and other cryptographic building blocks: A survey. *ACM Computing Surveys (CSUR)*, 55(2), 1–37.
8. Hossein, K. M., Esmaeili, M. E., Dargahi, T., Khonsari, A., & Conti, M. (2021). BCHealth: A novel blockchain-based privacy-preserving architecture for IoT healthcare applications. *Computer Communications*, 180, 31–47.

9. Kumar, P., Gupta, G. P., & Tripathi, R. (2021). TP2SF: A trustworthy privacy-preserving secured framework for sustainable smart cities by leveraging blockchain and machine learning. *Journal of Systems Architecture*, 115, 101954.
10. Zhang, E., Zhang, X., & Pareek, P. K. (2023). Herd effect analysis of stock market based on big data intelligent algorithm. In: Xu, Z., Alrabaee, S., Loyola-González, O., Cahyani, N. D. W., & Ab Rahman, N. H. (Eds.), *Cyber Security Intelligence and Analytics. CSIA 2023. Lecture Notes on Data Engineering and Communications Technologies* (Vol. 172). Springer. https://doi.org/10.1007/978-3-031-31860-3_14
11. Zhang, S., Yao, T., Arthur Sandor, V. K., Weng, T. H., Liang, W., & Su, J. (2021). A novel blockchain-based privacy-preserving framework for online social networks. *Connection Science*, 33(3), 555–575.
12. Qashlan, A., Nanda, P., He, X., & Mohanty, M. (2021). Privacy-preserving mechanism in smart home using blockchain. *IEEE Access*, 9, 103651–103669.
13. Soni, M., & Singh, D. K. (2021, June). Blockchain implementation for privacy preserving and securing the healthcare data. In: *2021 10th IEEE International Conference on Communication Systems and NETWORK Technologies (CSNT)* (pp. 729–734). India: IEEE.
14. Zhang, C., Zhao, M., Zhu, L., Zhang, W., Wu, T., & Ni, J. (2022). FRUIT: A blockchain-based efficient and privacy-preserving quality-aware incentive scheme. *IEEE Journal on Selected Areas in Communications*, 40(12), 3343–3357.
15. Singh, R., Dwivedi, A. D., Srivastava, G., Chatterjee, P., & Lin, J. C. W. (2023). A privacy preserving internet of things smart healthcare financial system. *IEEE Internet of Things Journal*, 10(21), 18452–18460.
16. Kumar, M., Mukherjee, P., Verma, S., Kavita, Shafi, J., Wozniak, M., & Ijaz, M. F. (2023). A smart privacy preserving framework for industrial IoT using hybrid meta-heuristic algorithm. *Scientific Reports*, 13(1), 5372.
17. Safaei Yaraziz, M., Jalili, A., Gheisari, M., & Liu, Y. (2023). Recent trends towards privacy-preservation in internet of things, its challenges and future directions. *IET Circuits, Devices & Systems*, 17(2), 53–61.
18. Bali, V., Bali, S., Gaur, D., Rani, S., & Kumar, R. (2023). Commercial-off-the shelf vendor selection: A multi-criteria decision-making approach using intuitionistic fuzzy sets and TOPSIS. *Operational Research in Engineering Sciences: Theory and Applications*, 6(2).
19. Kumar, P., Banerjee, K., Singhal, N., Kumar, A., Rani, S., Kumar, R., & Lavinia, C. A. (2022). Verifiable, secure mobile agent migration in healthcare systems using a polynomial-based threshold secret sharing scheme with a blowfish algorithm. *Sensors*, 22(22), 8620.
20. Rani, S., Kataria, A., & Chauhan, M. (2022). Cyber security techniques, architectures, and design. In: *Holistic Approach to Quantum Cryptography in Cyber Security* (pp. 41–66). USA, Germany: CRC Press.
21. Chauhan, M., & Rani, S. (2021). Covid-19: A revolution in the field of education in India. *Learning How to Learn Using Multimedia*, 23–42.
22. Rani, S., Pareek, P. K., Kaur, J., Chauhan, M., & Bhambri, P. (2023, February). Quantum machine learning in healthcare: Developments and challenges. In: *2023 IEEE International Conference on Integrated Circuits and Communication Systems (ICICACS)* (pp. 1–7). Raichur: IEEE.
23. Puri, V., Kataria, A., Solanki, V. K., & Rani, S. (2022, December). AI-based botnet attack classification and detection in IoT devices. In: *2022 IEEE International Conference on Machine Learning and Applied Network Technologies (ICMLANT)* (pp. 1–5). Salvador: IEEE.
24. Sudevan, S., Barwani, B., Al Maani, E., Rani, S., & Sivaraman, A. K. (2021). Impact of blended learning during Covid-19 in Sultanate of Oman. *Annals of the Romanian Society for Cell Biology*, 14978–14987.

25. Sharma, P., Namasudra, S., Chilamkurti, N., Kim, B. G., & Gonzalez Crespo, R. (2023). Blockchain-based privacy preservation for IoT-enabled healthcare system. *ACM Transactions on Sensor Networks*, 19(3), 1–17.
26. Kataria, A., Agrawal, D., Rani, S., Karar, V., & Chauhan, M. (2022). Prediction of blood screening parameters for preliminary analysis using neural networks. In: *Predictive Modeling in Biomedical Data Mining and Analysis* (pp. 157–169). USA: Academic Press.
27. Rani, S., Bhambri, P., & Kataria, A. (2023). Integration of IoT, big data, and cloud computing technologies. In: *Big Data, Cloud Computing and IoT: Tools and Applications*. USA: CRC Press.
28. Tanwar, R., Chhabra, Y., Rattan, P., & Rani, S. (2022, September). Blockchain in IoT networks for precision agriculture. In: *International Conference on Innovative Computing and Communications: Proceedings of ICICC 2022* (Vol. 2, pp. 137–147). Singapore: Springer Nature Singapore.
29. Bhambri, P., & Rani, S. (2024). Challenges, opportunities, and the future of industrial engineering with IoT and AI. *Integration of AI-Based Manufacturing and Industrial Engineering Systems with the Internet of Things*, 1–18.
30. Gao, Y., Chen, L., Han, J., Wu, G., & Susilo, W. (2023). IoT privacy-preserving data mining with dynamic incentive mechanism. *IEEE Internet of Things Journal*, 11(1), 777–790.
31. Bhambri, P., Rani, S., Balas, V. E., & Elngar, A. A. (Eds.). (2023). *Integration of AI-Based Manufacturing and Industrial Engineering Systems with the Internet of Things*. USA: CRC Press.
32. Rani, S., Kaur, J., & Bhambri, P. (2023). Technology and gender violence: Victimization model, consequences and measures. In: *Communication Technology and Gender Violence* (pp. 1–19). Cham: Springer International Publishing.
33. Singh, H., & Rani, S. (2023). Flex sensor integrated smart strap to verify correct wearing of the face mask. *IEEE Sensors Journal*, 24(2), 2020–2027.
34. Rani, S., Kumar, S., Kataria, A., & Min, H. (2023). SmartHealth: An intelligent framework to secure IoMT service applications using machine learning. *ICT Express*, 10(2), 420–427.
35. Gheisari, M., Javadpour, A., Gao, J., Abbasi, A. A., Pham, Q. V., & Liu, Y. (2023). PPDMIT: A lightweight architecture for privacy-preserving data aggregation in the internet of things. *Journal of Ambient Intelligence and Humanized Computing*, 14(5), 5211–5223.
36. Rani, S., Mishra, A. K., Kataria, A., Mallik, S., & Qin, H. (2023). Machine learning-based optimal crop selection system in smart agriculture. *Scientific Reports*, 13(1), 15997.
37. Kataria, A., Puri, V., Pareek, P. K., & Rani, S. (2023, July). Human activity classification using G-XGB. In: *2023 International Conference on Data Science and Network Security (ICDSNS)* (pp. 1–5). Raichur: IEEE.
38. Al-Shabandar, R., Jaddoa, A., Liatsis, P., & Hussain, A. J. (2021). A deep gated recurrent neural network for petroleum production forecasting. *Machine Learning with Applications*, 3, 100013.
39. Al-Abaji, M. A. (2021). Cuckoo search algorithm: Review and its application. *Tikrit Journal of Pure Science*, 26(2), 137–144.

8 Blockchain-Based Vehicle Information Management System

Anwesha Banik, Waseem Ahmad Mir, Tawseef Ayoub Shaikh and Iqra Nissar

8.1 INTRODUCTION

The invention of vehicles has made our lives easier and more flexible. For this reason, the automobile industry has grown so much that it is a leading factor in contributing to India's gross GDP. The Regional Transport Office (RTO) is the authority for registering vehicles and providing vehicle identification numbers, license numbers, etc. With the advent of the e-governance initiative, the Ministry of Road Transport Highways (MoRTH) launched VAHAN for registering vehicles and SARATHI for issuing driving licenses. These web applications improve the user experience as well as reduce the workload of RTOs. As the proverb says, each coin has two sides. Similarly, e-governance has some disadvantages. The government has made "data breeches" occur frequently. Data breaches are situations in which data has been lost or exposed to an unauthorized person. Data breeches mainly occur due to insider threats, weak login credentials, etc. The e-governance architecture is built upon the client-server architecture. The whole dataset is stored in a database from where it can be queried on client request. The data write or read access is given to employees for their job roles. Hence, there may be a chance of the addition of fake records to the database by dishonest employees for their personal profit. Fake records encourage social crimes such as vehicle robbery and can also mislead police investigations. Another drawback in the existing system is that different state RTOs and manufacturers are not bidden into one platform. Each entity does not hold complete information about a vehicle. For this reason, stolen vehicles or criminal activities involved vehicles can be re-registered into new states as new state RTOs are not aware of the past history. Moreover, e-governance made data an easy target for cyberattackers. Existing security mechanisms cannot tackle cyberattacks like denial of service, malicious code injection, cross-server scripting, etc. The security vulnerabilities can be mitigated using premised blockchain. Blockchain is a link list-based data structure that provides a decentralized, distributed ledger. Permissioned blockchain provides a decentralized ledger where each participant knows each other and can verify or validate the transaction.

DOI: 10.1201/9781003460367-8

8.2 BACKGROUND

This section describes the background of vehicle information management system in a systematic manner.

8.2.1 Cryptography Primitives

Public Key Cryptography: Cryptography is the branch of science that provides a mechanism for securing communication [1]. It ensures confidentiality, integrity and authentication of messages. Confidentiality ensures data is protected from unauthorized access. Data integrity means data is not altered before it is received at its destination. Authentication makes us aware that data has been destined to be valid receiver.

Encryption is a cryptographic technique where plain text is converted to cipher text [2–5]. The plain text is the data that has been sent over the network. Before being transmitted over a network, plain text is rearranged and mapped with other elements to form cipher text. Decryption is the conversion of cipher text back to plain text. There are two types of encryption–decryption techniques: asymmetric and symmetric (Figure 8.1). The symmetric key algorithms are the encryption–decryption algorithms where the same key is used at the sender and receiver ends to encrypt and decrypt the data, respectively. DES, 3DES, AES, IDEA, RC4, and RC5 are some

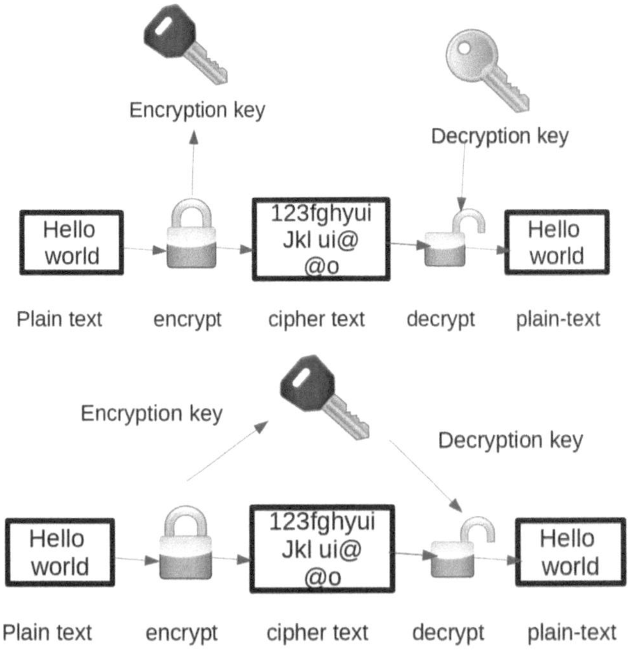

FIGURE 8.1 (a) Symmetric encryption. (b) Asymmetric encryption.

standard symmetric key algorithms. Asymmetric algorithms are encryption and decryption algorithms where different key pairs are used at the sender and receiver ends for encrypting and decrypting data. It is also known as public key cryptography. Diffie–Hellman and RSA are the two most popular algorithms in public key cryptography. Public key cryptography has a use case in blockchain. It is used to verify and validate the authenticity of people joining the blockchain network.

Hash Function: Hash function is a function that takes an arbitrary length string into fixed length output. The output is known as message digest or hash code [6].

A cryptographic hash function has the following properties:

(a) *One Way Function:* The property states that given the hash function (*H*) and given the hash value (h), it is infeasible to find an input *m* such that $H(m) = h$.
(b) *Target Collision Resistance:* For a given hash function *H*, it is infeasible to draw out input *m* and *m′* whose $H(m) = H(m')$.
(c) *Deterministic:* This property ensures that hash function *H* should be given same hash value.
(d) *Avalanche Effect:* It ensures changing one bit should change the output drastically.

Digital Signature: A digital signature provides a mechanism for validating the authenticity of a message (Figure 8.2) [7]. The digital signature mechanism starts when the sender wants to send a message to the intended recipient. The sender's message (M) is sent to a hash function, and the output of the hash function is encrypted using the sender's private key. The encrypted hash digest along with the original message are sent over the network. The receiver, upon receiving the encrypted hash digest along with the original message, decrypts the encrypted hash digest with the sender's public key. If the decrypted hash digest corresponds to the encrypted original message, then the receiver can verify whether the message has been altered during transmission or not [8–12].

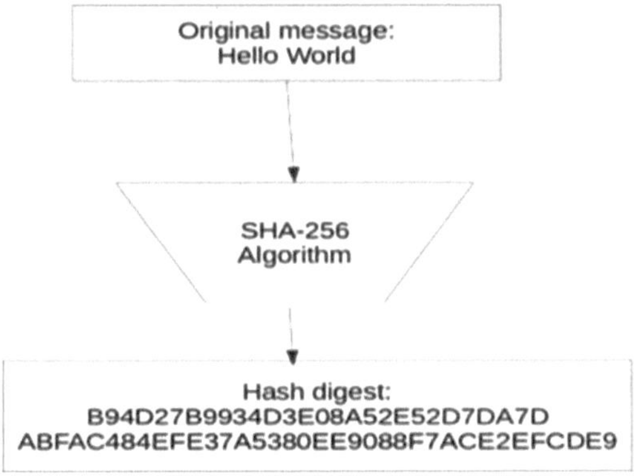

FIGURE 8.2 Hash function for an input m always.

8.2.2 Blockchain Architecture

A blockchain is a tamper-proof, link list-structured ledger in which data that has been added cannot be deleted or modified. It has a decentralization property that allows everyone to keep a copy of the ledger and process the data locally. Any information that is stored in a blockchain is known as a transaction. A block is a batch of such validated transactions. The ledger grows with the addition of new blocks of data. It also ensures that every peer on the network has the same copy of the ledger. Everyone can verify or validate the transaction, as the ledger is transparent to everyone on the network. A hashing technique is used on the blockchain to achieve immutability [13–17]. The entire block of data, along with the randomly generated number, is hashed and stored on the next block, thus forming a chain of blocks. If someone modifies a value in a block, the adversary needs to change the hash of the next corresponding block, which is computationally infeasible.

8.2.3 Block

A block in a blockchain has two parts: a block header and a block body.

8.2.3.1 Block Header

A block header is an 80-byte long string which consists of five fields.

Block-Version: The 4-byte long field gives information about what version of the block is being used. There are different versions of the block that specify which rules to be followed for block validation.

Merle-Tree-Root-Hash: It is a 32-byte-long field which consists of the hash of hassles of all transactions present in that particular block.

Difficulty Target: It is a 256-bit field that specifies a threshold for miners. The SHA-256 of the block header plus the nonce value should be less than or equal to the target provided.

Nonce: The 4-byte field, which starts with zero and gradually increases with the increase of each transaction. The combined hash of the nonce and the block header forms the hash of the previous block.

Parent Block Hash: A 256-bit hash value points to previous block.

8.2.3.2 Block Body

A transaction counter and transactions make up the block body. The maximum number of transactions that a block can hold depends upon the block size and the size of each transaction (Figure 8.3).

8.2.4 Characteristics of Blockading

The characteristics of blockchain are as follows:

(a) *Decentralization:* Decentralization is the most important property of the blockchain. Due to the decentralized nature of blockchain, no single entity has the power to modify or seize control of the entire ledger. Each peer has

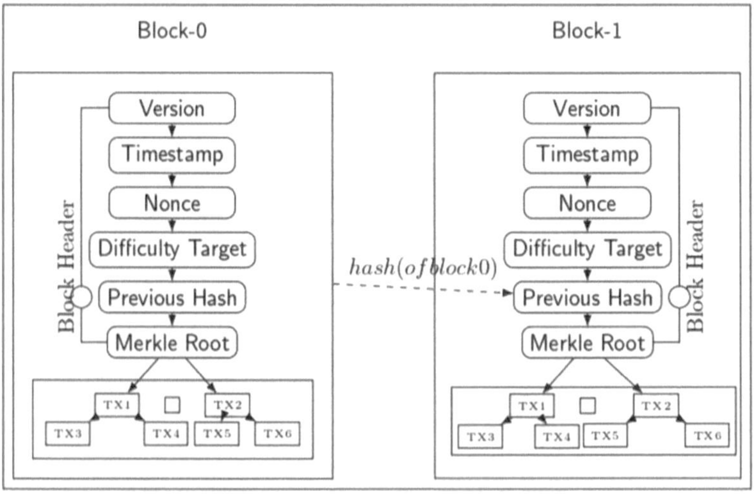

FIGURE 8.3 Blockchain architecture.

a copy of the ledger, and when a transaction needs to be added, everyone decides together.

(b) *Immutability:* The immutability of Blockchain can be achieved by using a cryptographic hash function. Each element is independently hashed and jointly hashed with a random number to form the hash of the entire block, which is stored in subsequential blocks. If any block data is modified, then the adversary needs to change the hash of every block in the chain, which demands high computational resources and time, which is practically infeasible [18–20].

(c) *Transparency:* The Blockchain ledger is replicated and distributed among all network participants. Every peer is aware of all network transactions and can independently verify or validate each transaction. If any transaction receives a majority of approval from peers, the transaction is added to the block.

(d) *Non-Repudiation:* Non-repudiation means that no sender or receiver can deny the validity of a transaction upon sending or receiving it, respectively. All the transactions are recorded on blockchain and the data cannot be modified after being stored, hence later no party can deny the transaction.

(e) *Message Authentication:* An encrypted hash digest along with the original message is sent to the destination. The receiver checks the authentication of the message by matching the decrypted hash code with the hash of the original message. If there is a match, then it proves the integrity of the message.

(f) *Traceability:* Blockchain provides easy tracking of records, as the ledger can be viewed at any lace or anytime.

8.2.5 Consensus in Blockchain

Consensus provides a mechanism through which peers validate each transaction and reach to a common agreement. Through consensus mechanism, new block is created or deleted. The peers in the network who add block are known as miners. There are various consensus algorithm and they differ based on type of blockchain platform we are using. A blockchain can be permissioned or permissionless. In permissioned blockchain, consensus is achieved among known and pre-authenticated peers. Whereas in permissionless blockchain, anyone can participate in consensus mechanism. The following are the types of consensus protocol used in permissionless blockchain:

(a) **Proof of Work**
Proof of Work is also known as Nakamoto Consensus [21]. Proof of Work comes into existence with emergence of Bitcoin in 2009 by Satoshi Nakamoto. In Proof of Work, miners needed to solve a computationally challenging problem in exchange of incentives. The Proof of Work puzzle is to find a nonce value such that when hashed with block parameters such as Merkle tree, previous block hash gives a value which is less than target block hash. Once such nonce is found, miners create block and distribute across the network peers; the rest of the peers can verify easily by computing hash of the block and checking whether it satisfies the condition to be smaller than the current target value. The Proof of Work suffers from selfish mining [22]. It is possible for two miners to find nonce for a particular block at the same time, which makes forking of blockchain. It is necessary to propagate the block mining information to nearby peers. Selfish miners hide the information about newly mined block from other peers until the side chain become larger than the main chain. In this way, the hashing power of honest miners got wasted. Proof of Work involves wastage of valuable resources like money, energy, space, etc. It is estimated that by 2018, 0.3% of whole electricity will be spent on verifying transaction. A flowchart of PoW is shown in Figure 8.4.

(b) **Proof of Stake**
Proof of Stake is implemented in order to avoid the resource wastage in Proof of Work. Here, leaders are elected based on the stakes of an individual in the network. By stake, we mean the number of digital tokens that a particular node is holding. Some Proof of Stake blockchains, such as Cardano, use the Follow-the-Satoshi (FTS) algorithm to elect leaders. The FTS algorithm was proposed by Li Qiwei in 2012 [23]. FTS takes an arbitrarily length string like block hash as input and outputs the token index. Using the token index, the transaction history is searched and the present owner of the token is found. Consensus in Permissioned blockchain provides a mechanism through which a common agreement can be reached in presence of crash fault or Byzantine fault. A node is termed as Byzantine if it propagates wrong information in the network. Crash fault occurs when node stops operation due to software or hardware faults.

(c) **Paxos**
Paxos is a crash-fault-tolerant distributed consensus algorithm in asynchronous environment [24]. It is based on state machine replication. There are three actors in this algorithm: Proposer, Acceptor, and Learner. Proposer is

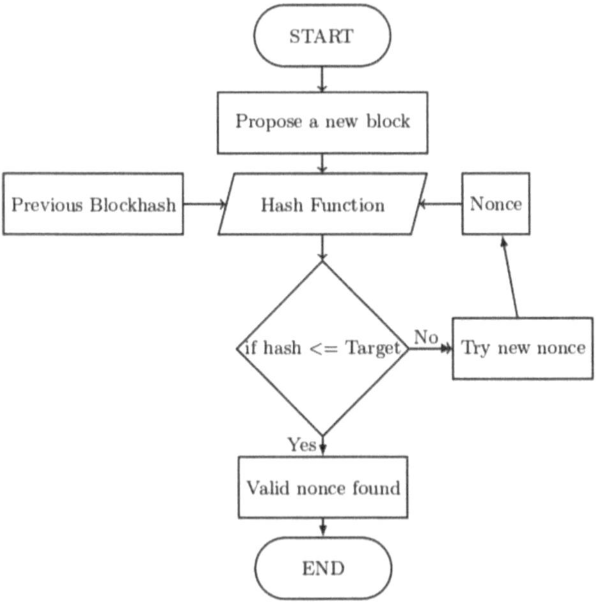

FIGURE 8.4 Proof of Work flowchart.

the one who proposes a value to all acceptors on behalf of clients. Acceptor is the one who accepts proposal. Learner is the one who learns the current proposal. Steps in Paxos algorithm are as follows:

I: Proposer Send PREPARE *IDP message* to all acceptors. I_{DPi} the unique identification mark of all PREPARE message.

II: Acceptors on receiving PREPARE *IDP message*, checks whether they promised to ignore this value or not. If not, then send Reply with PROMISE I_{DP}.

III: If Proposer gets majority of PROMISE messages for a specific *IDP*, then proposer Sends ACCEPT–REQUEST *IDP*, VALUE to a majority (or all) of Acceptors.

IV: Acceptor receives an ACCEPT-REQUEST I_{DP} and check whether they earlier promised to ignore with this *IDP*. *IF* yes, then they ignore. Otherwise, they Reply with ACCEPT I_{DP}, VALUE; Also send it to all learners.

V: If a proposer/learner gets majority of accept for a e-governance initiative, which provides a flexible interface for facilitating online RTO services like registration of vehicles, issuing driving licenses, etc." Vahan" is a repository of digital vehicle records. MoRTH also came up with a mobile application known as mParivahan, through which one can get instant access to vehicle information with just specific *IDP*, they know that consensus is reached for one click. In the case of the offline mode of new vehicle registration, one needs to visit the nearest RTO office and the value (not *IDP*).

8.3 CLASSIFICATION OF BLOCKCHAIN

(a) *Public Blockchain:* It is a type of blockchain where anyone can join the network. The participants may know or may not know each other. All the people in the network can verify or validate a new piece of information. The real identities of the participants are hidden. A Proof of Work mechanism is used for conducting condense. Bitcoin and Ethereum are such public blockchain.

(b) *Private Blockchain:* It is a type of permissioned blockchain where participants know each other but do not trust each other. The key credentials are required to log into the blockchain. Many permissioned blockchains, such as Hyperledger Fabric, allow you to specify who is authorized to verify or validate a transaction. Permissioned blockchain are best suited for business applications.

8.4 EXISTING VEHICLE REGISTRATION SYSTEM

It is compulsory to register his/her own vehicle with the Regional Transport Office to ply on Indian road. RTO has various responsibilities:

I. RTO collects road tax that an individual need to pay while registering the vehicle. Dual need to pay while registering the vehicle.
II. RTO provides vehicle identification number, driving license which are necessary documents for driving on the road.
III. RTO also issues permit for commercial vehicles.

Earlier, the management of vehicle data was based on files. Then, with the advancement of technology, RTO became computerized. But, with the inflation in the sales of motor vehicles, there is an increase in the workload on the RTO. It is necessary for human–machine interface for the smooth functioning of the department. For this reason, the Ministry of Road Transport and Highways (MoRTH) launched a "VAHAN" web portal [25], requiring to fill an application with necessary documents like valid identity proof, purchase invoice for the vehicle, copy of vehicle insurance, PAN Card copy, etc. After proper verification and validation, a unique vehicle identification number is allotted to the owner and the record is stored in their own database. Ensuring integrity–confidentiality–security for vehicle data has become challenging as existing system surfers from both internal and external threats. One of the principal reasons for presence of malicious activities in existing system is that records are stored in a database system. Database system poses certain security threats which can expose confidential vehicle data to outsiders. A database is a collection of useful facts and numbers that need to be stored. DBMS is a computerized record-keeping system which make data manipulation easy. It stores data in the form of tables. It supports multiple views of database. The data in which particular group of users have authority to query are shown and rest of the data is restricted for the user to manipulate. The characteristics of DBMS make it a popular technology for organization to store vulnerable data.

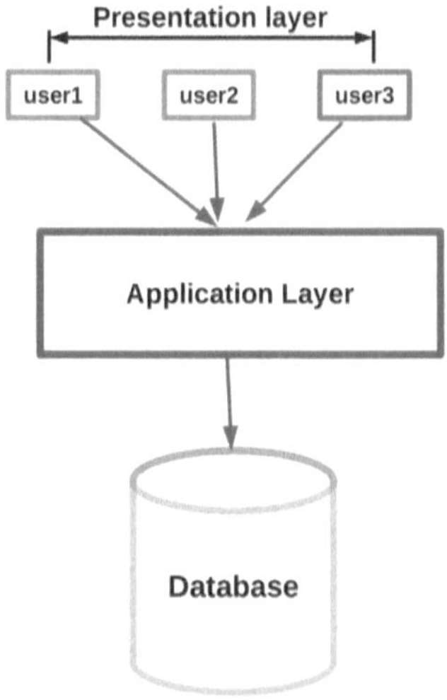

FIGURE 8.5 A typical three-tier architecture.

8.4.1 Typical Architecture for an e-Governance Application

A typical e-Governance Portal is of Tier-3 architecture [26]. Figure 8.5 depicts a three-tier architecture. Data backup, recovery, security, and concurrency control are characteristics of the three-tier architecture. The following are the layers of three-tier architecture:

1. *Presentation Layer:* It is also known as the "client layer." It provides a web interface to take user requests and display necessary information in the form of static or dynamic web pages.
2. *Application Layer:* It is also known as the "business–logic layer." It acts as a middleware between the client layer and the application layer. It consists of predefined logic which checks user credentials before forwarding to web servers.
3. *Database Layer:* Here in this layer, actual data is stored. It provides operation like insertion–deletion–updation for connection with the database.

8.4.2 Security Threats in DBMS and Mitigation through Blockchain

With advantages, the DBMS also has shortcomings regarding security. Security mechanism is important for DBMS in order to protect data from unprivileged use and data loss. The security mechanism must ensure confidentiality, integrity, and availability. Confidentiality ensures data must be protected from outsiders, Integrity confirms that data need to be manipulated by authorized peers. Availability facilities data need to be available to the users all the time. Some of the threats are discussed in following sections:

(a) *Data Breach:* Data breach is the violation of security measures where data is stolen or exposed to unauthorized person without intervention of valid–authorized person. The Data breach can also be teamed as data-leakage or data-spill. It is very necessary to avoid data breach as it causes financial loss as well as diminish business growth. In India, there is about 377% increase in data breach. Data breach is an outcome of internal and external threats. Internal threats occur due to legitimate privilege abuse. External breaches occur when an organization deploys a third party for a particular function. Mitigation using blockchain data leakage is not possible in blockchain as blockchain is append only ledger where ledger is distributed across all the peers. Each node maintains its own block of information. If an attacker has gained aces to one block, it is practically impossible to gain aces to subsequent blocks as it demands high computational power.

(b) *Data Transparency:* The data are stored in a single computer which is known as centralized database system. It is under the control of any single organization. Any adversary on public records done by the authorized organization cannot be detected as they are the sole proprietor of data. For example, in case of vehicle registration, RTO is the sole authority of vehicle data. Data modification, issuing driving license to fake person, and issuing fake paper proof as valid are the adversaries that can happen. The nodes which join the network may be located in different parts of the world. The blockchain ledger is replicated across the peers. Nodes collectively take decision through consensus mechanism to add or delete a transaction. Hence, there is no central authority present in blockchain who is the sole proprietor of data.

(c) *Data Mutability in Database:* The data which are stored in a single computer is known as centralized database system. It is under the control of any single organization. Any adversary on public records done by the authorized organization cannot be detected as they are the sole proprietor of data. For example, in case of vehicle registration, RTO is the sole authority of vehicle data. Data modification, issuing driving license to fake person, issuing fake paper proof as valid are the adversaries can happen. Data mutability is not possible in blockchain. Block Data is hashed along with nonce and they are again hashed with block header to form hash of a block and it is stored in next corresponding block. If any adversary tries to modify any data, he needs to modify subsequent block which is practically not possible as huge amount of computation power is needed.

8.5 NETWORK SECURITY THREATS AND MITIGATION USING BLOCKCHAIN

Denial-of-Service Attack: A denial-of-service attack is a type of attack where an adversary aims to consume all the network resources like bandwidth or try to make server down in an ill-intention to make legitimate users deprived of services. There

TABLE 8.1
Threats in DBMS and Mitigation with Blockchain

DBMS Threats	Solution/Blockchain
Data breach	Loss of information is not possible as ledger is replicated across all the nodes
Data mutability	Hash of block is stored in next corresponding block which make the blockchain tamper-proof
Data opacity	As blockchain is decentralized, data is transparent to all the peers joining in the network

are three types of denial-of-service attacks. Various DBMS threats and blockchain mitigation methods are shown in Table 8.1.

(a) *SYN Flooding Attack:* In this attack, adversaries initialize connection with server but do not finalize the connection. The connection remains half-opened and server needs to wait and spends resources on it which can eventually drain all the resources of a server.
(b) *UDP Flooding Attack:* UDP flooding attack is like SYN flooding attack, where the target server is requested with a large number of UDP packets in such a manner that it freezes the machine ability to respond and report.
(c) *Distributed Denial-of-Service (DDoS) Attack:* When a DoS attack is performed to a particular server from multiple distributed sources, it is known as distributed denial-of-service attack.

When a malicious node in a blockchain network transmits a high number of transactions with the goal of slowing down processing power, it is known as a denial-of-service attack. By restricting the number of transactions that can be sent by a node per second, the denial-of-service attack can be lessened. Denial-of-service can also be prevented by disconnecting the node that sends out too many transactions.

SQL Injection: DBMS provides Structural Query language (SQL) to interact with the database. By querying, we can insert, delete, or modify record in database. SQL injection attack happens when a malicious individual inputs a value, which in turn translated to a query statement, which can retrieve hidden information from database. Example of SQL Injection:

Suppose we have an application called e-transport, where all vehicle records are stored. Each vehicle has an individual vehicle identification number. Let, for instance, we want to retrieve status of vehicle record with VIN "1234." If we input VIN number as 1234 on API, it basically translates into SQL query.

*SELECT * from TRANSPORT where VIN == 1234*
But if we input VIN as 1234 or 1 = 1 in the input field, it gets translated to:
*SELECT * from TRANSPORT where VIN == 1234 or 1 = 1*
The "or 1 = 1" appended with input retrieve all records where condition is true.
By this, a malicious user can have grasp over whole TRANSPORT data by simply manipulating SQL query.

SQL Injection: Diminishing Scheme Using Blockchain: This kind of attack can be mitigated by using permissive blockchain. Here, participants know each other and they all jointly maintain a ledger to fulfill their purpose, and all the information is transparent to each one of them, so there is no need for any one of them to know any information by injecting malicious code. If someone tries to update the smart contract using malicious code, it will be sent as a transaction to all the peers for approving the smart contract update. If the smart contract update is malicious, it would not be approved by the majority of the participants, and the code injector can be identified easily.

8.6 PERMISSIONED BLOCKCHAIN: A BETTER SOLUTION FOR VEHICLE REGISTRATION SYSTEM

The advantages of using a permissive blockchain are explained here. The permissioned blockchain allows known identities to join the network. Some permissioned blockchains, such as Hyperledger Fabric, have endorsement policies that allow you to configure which peers have read or write access. This type of blockchain enables two or more organizations to share a common ledger. It works on an execute–order–validate mechanism. It is available with smart contracts, where logic is written that needs to be satisfied for each transaction to get into the network. Permissioned blockchain is appropriate for vehicle registration systems because only authorized individuals can join the network, and it is easily deployable due to its low computing power requirements. The probability of hacking attacks, replay attacks, and malicious code injection is decreased. The regional transport office, manufacturer, and police can be bound into one platform using permissioned blockchain, which will obsolete the false injection of records, as well as re-registration of stolen vehicles. The data immutability and decentralization properties of blockchain prevents legitimate user to abuse their access rights.

8.7 CONCLUSION

In this chapter, the issues present in cutting-edge vehicle data management system are illuminated. The existing technique suffers from ill-use of admissible rights. There are also presence of some external threats like false SQL query which can make data privacy to question. Enforcement of security–confidentially–privacy of vehicle data can be achieved through blockchain. Blockchain is link-list-based immutable ledger which can address all the menace of existing system. Different techniques based on blockchain provided by different authors have been discussed here.

REFERENCES

1. Mushtaq, M. F., Jamel, S., Disina, A.-D. H., Pindar, Z. A., Ahmad Shakir, N. S., & Mat Deris, M. (2017). A survey on the cryptographic encryption algorithms. *International Journal of Advanced Computer Science and Applications*, 8(11).
2. Bali, V., Bali, S., Gaur, D., Rani, S., & Kumar, R. (2023). Commercial-off-the shelf vendor selection: A multi-criteria decision-making approach using intuitionistic fuzzy sets and TOPSIS. *Operational Research in Engineering Sciences: Theory and Applications*, 6(2).

3. Kumar, P., Banerjee, K., Singhal, N., Kumar, A., Rani, S., Kumar, R., & Lavinia, C. A. (2022). Verifiable, secure mobile agent migration in healthcare systems using a polynomial-based threshold secret sharing scheme with a blowfish algorithm. *Sensors*, 22(22): 8620.
4. Rani, S., Kataria, A., & Chauhan, M. (2022). Cyber security techniques, architectures, and design. In *Holistic Approach to Quantum Cryptography in Cyber Security* (pp. 41–66). USA: CRC Press.
5 Chauhan, M., & Rani, S. (2021). Covid-19: A revolution in the field of education in India. *Learning How to Learn Using Multimedia*: 23–42.
6. Sobti, R., & Geetha, G. (2012). Cryptographic hash functions: A review. *International Journal of Computer Science Issues (IJCSI)*, 9(2): 461.
7. Matyas, S. M. (1979). Digital signatures—an overview. *Computer Networks (1976)*, 3(2): 87–94.
8. Rani, S., Pareek, P. K., Kaur, J., Chauhan, M., & Bhambri, P. (2023, February). Quantum machine learning in healthcare: Developments and challenges. In *2023 IEEE International Conference on Integrated Circuits and Communication Systems (ICICACS)* (pp. 1–7). Raichur: IEEE.
9. Puri, V., Kataria, A., Solanki, V. K., & Rani, S. (2022, December). AI-based botnet attack classification and detection in IoT devices. In *2022 IEEE International Conference on Machine Learning and Applied Network Technologies (ICMLANT)* (pp. 1–5). Salvador: IEEE.
10. Sudevan, S., Barwani, B., Al Maani, E., Rani, S., & Sivaraman, A. K. (2021). Impact of blended learning during Covid-19 in Sultanate of Oman. *Annals of the Romanian Society for Cell Biology*: 14978–14987.
11. Kataria, A., Agrawal, D., Rani, S., Karar, V., & Chauhan, M. (2022). Prediction of blood screening parameters for preliminary analysis using neural networks. In *Predictive Modeling in Biomedical Data Mining and Analysis* (pp. 157–169). USA: Academic Press.
12. Rani, S., Bhambri, P., & Kataria, A. (2023). Integration of IoT, big data, and cloud computing technologies. In *Big Data, Cloud Computing and IoT: Tools and Applications*. USA: CRC Press.
13. Tanwar, R., Chhabra, Y., Rattan, P., & Rani, S. (2022, September). Blockchain in IoT networks for precision agriculture. In *International Conference on Innovative Computing and Communications: Proceedings of ICICC 2022* (Vol. 2, pp. 137–147). Singapore: Springer Nature Singapore.
14. Bhambri, P., & Rani, S. (2024). Challenges, opportunities, and the future of industrial engineering with IoT and AI. *Integration of AI-Based Manufacturing and Industrial Engineering Systems with the Internet of Things*: 1–18.
15. Bhambri, P., Rani, S., Balas, V. E., & Elngar, A. A. (Eds.). (2023). *Integration of AI-Based Manufacturing and Industrial Engineering Systems with the Internet of Things*. USA: CRC Press.
16. Rani, S., Kaur, J., & Bhambri, P. (2023). Technology and gender violence: Victimization model, consequences and measures. In *Communication Technology and Gender Violence* (pp. 1–19). Cham: Springer International Publishing.
17. Singh, H., & Rani, S. (2023). Flex sensor integrated smart strap to verify correct wearing of the face mask. *IEEE Sensors Journal*, 24(2): 2020–2027.
18. Rani, S., Kumar, S., Kataria, A., & Min, H. (2023). SmartHealth: an intelligent framework to secure IoMT service applications using machine learning. *ICT Express*, 10(2): 420–425.
19. Rani, S., Mishra, A. K., Kataria, A., Mallik, S., & Qin, H. (2023). Machine learning-based optimal crop selection system in smart agriculture. *Scientific Reports*, 13(1): 15997.
20. Kataria, A., Puri, V., Pareek, P. K., & Rani, S. (2023, July). Human activity classification using G-XGB. In *2023 International Conference on Data Science and Network Security (ICDSNS)* (pp. 1–5). Tiptur: IEEE.

21. Nakamoto, S. (2008). Bitcoin: A peer-to-peer electronic cash system. *Decentralized Business Review*, 21260.
22. Schwarz-Schilling, C., Li, S.-N., & Tessone, C.-D. J. (2022). Stochastic modelling of selfish mining in proof-of-work protocols. *Journal of Cybersecurity and Privacy*, 2(2): 292–310.
23. Nguyen, C. T., Hoang, D. T., Nguyen, D. N., Niyato, D., Nguyen, H. T., & Dutkiewicz, E. (2019). Proof-of-stake consensus mechanisms for future blockchain networks: Fundamentals, applications and opportunities. *IEEE Access*, 7: 85727–85745.
24. Lamport, L. (2001). Paxos made simple. *ACM SIGACT News (Distributed Computing Column) 32, 4 (Whole Number 121, December 2001)*, 51–58.
25. https://morth.nic.in/.
26. Kumar, M., Shukla, M., Agarwal, S., & Pandey, G. N. (2013). An e-governance model using cloud computing technology for developing countries. In *Proceedings of the International Conference on e-Learning, e-Business, Enterprise Information Systems, and e-Government (EEE)*, 171, India.

9 A Survey on Issues in Integrating Blockchain and IoT Technologies

Bharati B. Pannyagol and Santosh L. Deshpande

9.1 INTRODUCTION

Today, the most fundamental resource after food and shelter is probably the internet. If we are not connected to the internet, we face difficulties in making banking transactions, vocational plan, or self-learning. A groundbreaking technology called the Internet of Things (IoT) has evolved in the online world. It connects real-world systems, objects, and gadgets to the internet, and so they can communicate, share information, and act independently. It can function in harsh environmental circumstances with fewer resources. Additionally, it handles data processing and aggregation at both the network edge level and the cloud side. IoT is changing the way we behave and respond. Today we control air conditioners with our smartphone and our intelligent cars show us the shortest route—these are examples of smart technologies that monitor our daily activities. IoT signifies a vast network of connected items. These things collect and disseminate data regarding their use and the environment in which they are operated. It is all done using sensors. Every physical object contains sensors. Our mobile phone, electrical devices, Pecos barcode sensors, traffic lights, and nearly everything else that we encounter every day all can be affected (Kortuem et al., 2010). The essential question is how these sensors exchange this enormous amount of data and how we might use it for our benefit. These sensors continuously produce data about the devices' operational status (Bali et al., 2023). These gadgets may all connect with one another and dump their data onto a single platform with the help of the IoT. Data from various sensors is released, and IoT platform security receives it. IoT platforms incorporate the data gathered from numerous sources. The data is subjected to additional analysis, and useful information is extracted in accordance with needs. Other devices are made aware of the outcome to improve user automation. However, it faces many challenges due to its resource constraint such as size, battery life, etc. In different application areas that pertain to people's daily lives, the IoT revolutionizes commonplace objects to improve people's lives (Alam, 2018; Kortuem et al., 2010), including the environment, healthcare (Pal et al., 2016), agriculture, transportation (Ayoub et al., 2020), smart homes (Laouiti et al., 2020; Bae & Chang, 2012) and smart cities (Harrison et al., 2010; Kim et al., 2017), etc.

DOI: 10.1201/9781003460367-9

9.1.1 History of IoT (Sharma et al., 2019)

- In the late 1980s, data transmission was used for linking devices and enabling communication between them. The first internet-enabled device was invented at Carnegie Mellon University, which was a Coca Cola vending machine (Kumar et al., 2022).
- During 1990–2000, RFID technology was emerged. The RFID tags allow objects to be uniquely identifiable and it can easily track the devices. In the year 2000, wireless communication, sensor networks, and embedded networks all were introduced.
- In 1999, the term Internet of Thing was officially announced by researcher Kevin Ashton. Businesses adopt the IoT concept. The idea of smart homes and smart cities is beginning to take shape.
- In 2010, the organization International Telecommunications and IEEE started to define the protocols for IoT devices.
- The IoT network will connect a huge number of connected devices between 2010 and 2020. They begin to expand across a variety of industries, including healthcare, agriculture, transportation, business, and big data analytics and cloud computing start using the information gathered from IoT devices.
- As the number of IoT-connected devices grows in 2020, edge computing, AI, and blockchain technologies will be combined with IoT to speed up data processing, enable self-decision making, and offer security at the source or sink level.

9.1.2 Architecture of IoT

The IoT architecture (Gubbi et al., 2013; Sobin, 2020) refers to the structure and system design (Figure 9.1). It enables data transfer between various Internet-connected devices (Sobin, 2020; Al-Fuqaha et al., 2015).

9.1.2.1 Perception Layer

In the architecture, it is the base layer. There are sensors, actuators, and other data acquisition components in it. These devices collect the environmental data, real space, and object domain data and send it to the central devices. This layer is in charge of processing the data before it is transferred to the central device. The actuation is performed in this layer.

9.1.2.2 Network Layer

Between the physical layer and the application layer, this layer manages data flow and communication. It consists of gateway, network protocol, and connecting technologies. The gadgets are connected using both wired and wireless technologies.

9.1.2.3 Application Layer

It carries out data analytics using the information gathered from the perceptual layer. Through applications and user interfaces, it provided the decision that was made as a result of data processing. These interfaces make user to interact with IoT networks, environment monitoring, and real-time parameter setting.

FIGURE 9.1 IoT architecture.

The number of IoT devices worldwide is expected to reach 75.44 billion by the end of 2025, according to various industry estimates (Alam, 2018). Figure 9.2 shows the number of active IoT devices connected worldwide is larger compared to non-IT devices. Figure 9.3 illustrates the data collection by different IoT devices.

9.1.3 Issues and Challenges in IoT

We have surveyed different issues and challenges that occurred in IoT network (Atzori et al., 2010; Miorandi et al., 2012; Pannayagol & Deshpande, 2023).

- *Security and Privacy:* As there are a huge number of connected gadgets, security and privacy are two critical issues. Weak security methods can make the IoT network victim for hacking, data breaches, and unauthorized access. As massive amount of personal data is continuously gathered from different IoT devices, privacy concerns exist (Rani et al., 2022).
- *Interoperability:* The IoT network consists of the large amount of devices from different vendors, using different communication technologies and protocols. Achieving the common communication standards between the devices is a major issue (Chauhan & Rani, 2021).
- *Scalability:* As the number of devices is ever increasing, handling them is a very big issue. IoT has to handle large amount of data, process it, and

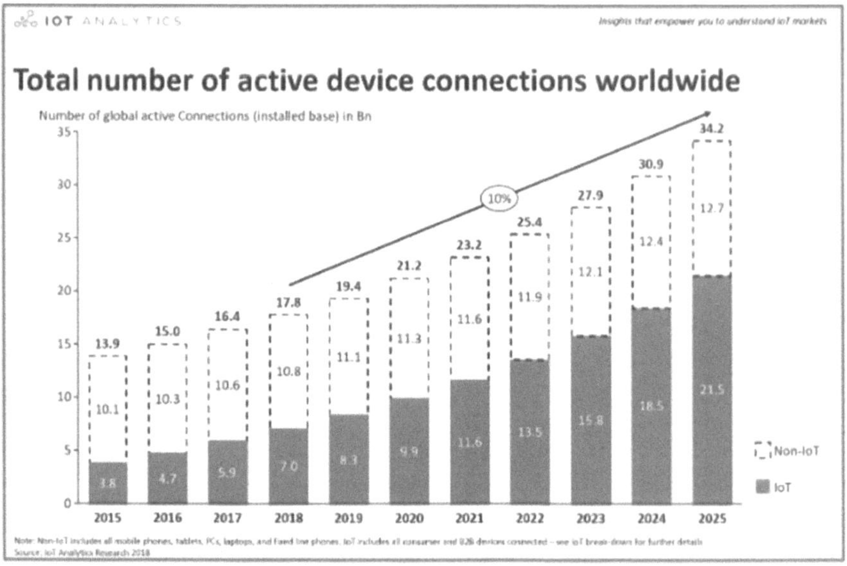

FIGURE 9.2 Total number of active device connections worldwide.

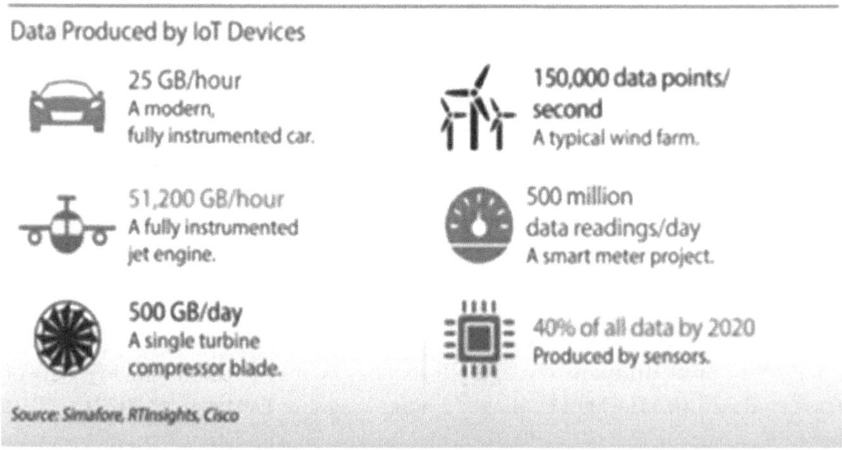

FIGURE 9.3 Data collected by different IoT applications.

analyze it properly to get accurate results. Providing good performance, low latency, and scalability are the challenges, especially in limited network bandwidth or processing power applications.

- *Power and Energy Efficiency:* Maximum number of IoT devices work on batteries. In particular, for remote or inaccessible deployments where regular battery replacements are impractical, balancing functionality, data transmission, and power efficiency is critical to extending the lifespan of IoT devices.

- *Infrastructure and Connectivity:* To support the connectivity, data transmission, and processing needs of IoT, a reliable and expandable infrastructure is necessary.

9.2 BLOCKCHAIN TECHNOLOGY

Blockchain (Meunier, 2018; Swan, 2015) is an append only shared ledger of digitally signed and encrypted transaction replicated across a network of peer nodes. Decentralization, immutability, and transparency are blockchain's primary characteristics. It uses a dispersed network to operate, to increase security and robustness. Data recorded on the blockchain cannot be tampered, promoting the trust among users. Smart contract automates operations and its transparency enables the audit ability. Transactions are fast and cost-effective. Peer-to-peer communication eliminate intermediaries. Users will protect their privacy and control the access of their data. The systems fault tolerance ensures that it will still continue to operate if some nodes fail or are compromised. Interoperability feature allows the seamless data exchange and collaboration among the different blockchain networks (Zheng et al., 2018).

9.2.1 Types of Blockchain

There are three main types of blockchain:

- *Public Blockchain:* It operates in a decentralized manner. It allows anyone to participate in creating new blocks and validate transaction. It provides the security by consensus mechanisms like Proof of Work (PoW) or Proof of Stake (PoS). It allows anybody to participate in the decision-making process. Because the network is public, creating new blocks needs a lot of processing power, which results in high transaction fees. Example: Ethereum and Bitcoin.
- *Private Blockchain:* In this, each participant is initially known. It can be used with consortiums or organization where a predetermined set of nodes or entities perform the network's management and validation. It provides higher privacy and faster transaction by restricting access to specific participants. Example: Hyperledger Fabric, R3 Corda.
- *Hybrid Blockchain:* It is a combination of private and public blockchains. By this strategy, we can achieve transparency and security by public blockchains while keeping the privacy of sensitive data. Records of data flows between consortium members are stored, cost of computing blocks and publishing them is inexpensive. These blockchains are frequently used in consortium like banks where it is required for numerous separate entities to collaborate. Example: Dragon chain, Aion (Sudevan et al., 2021).

9.2.2 Applications of Blockchain

Figure 9.4 illustrates the various blockchain applications. Due to its decentralized, secure, and transparent nature, several companies and sectors could be totally transformed by blockchain technology (Rani, Kataria, et al., 2023).

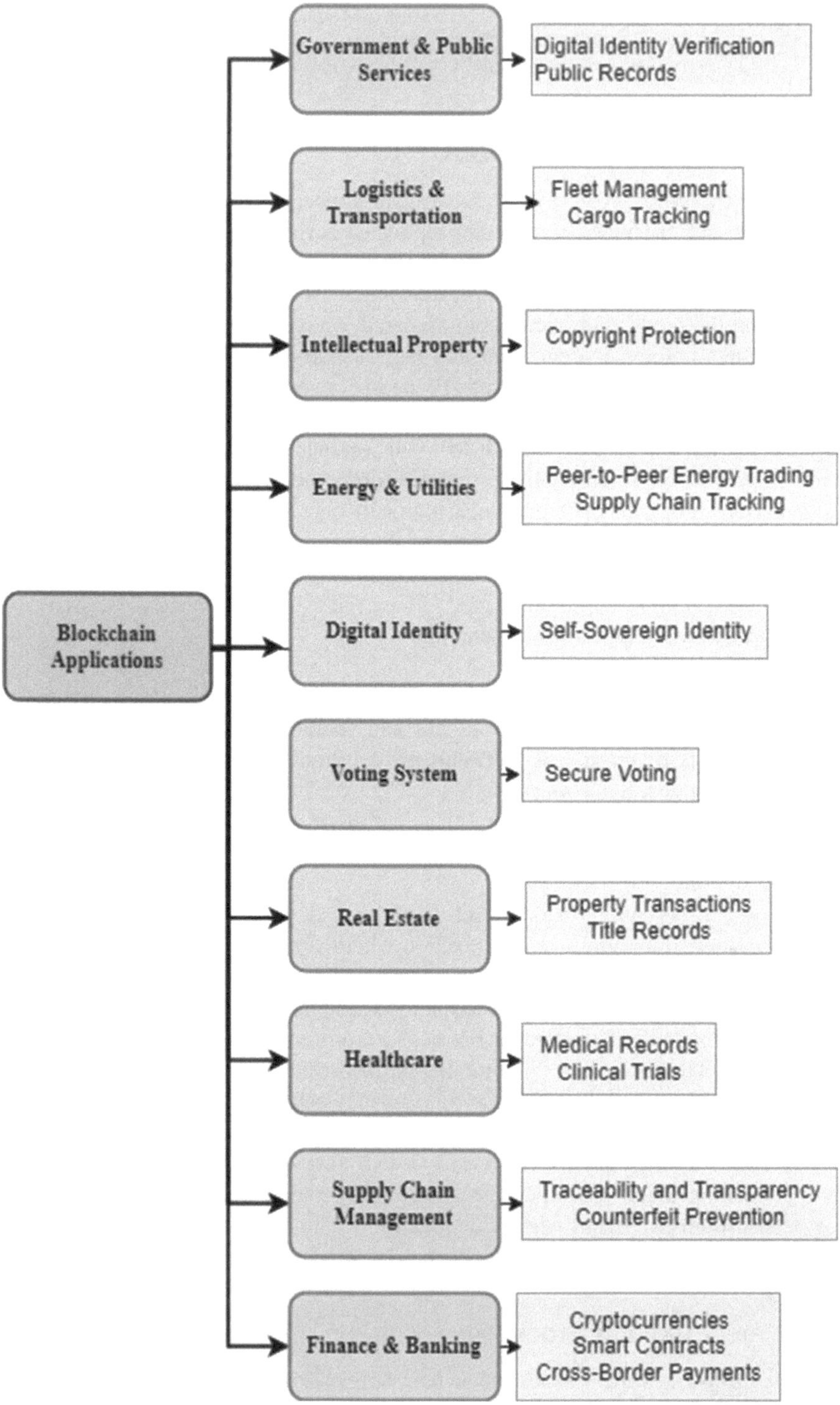

FIGURE 9.4 Applications of blockchain.

9.2.3 Blockchain Elements

We are going to see in detail the elements of blockchain (Zhang et al., 2023): First we will study the basic structure of block and then how the block will be processed in the blockchain (Tanwar et al., 2022, September).

9.2.3.1 Block Structure

Block: A block is a container data structure that contains a series of transactions. It is a digitally signed and encrypted set of transactions which are verified by peers (Rani, Kaur, et al., 2023).

Each block in the blockchain contains the following elements. Figure 9.5 depicts the structure of block in the blockchain.

- *Block Header:* It contains the following essential information about the block:
  - *Version:* The version of the block.
  - *Previous Block Hash:* In blockchain, every block inherits from previous block. This field contains the previous block headers hash value. With this field, the blockchain is made tamper-proof (Rani, Mishra, et al., 2023).
  - *Timestamp:* It indicates the time of the block creation or mined.
- *Merkle Root Hash:* It is a cryptographic hash value of all transactions included in a block. It guarantees that every transaction is securely and effectively connected to the block.

A Merkle tree is a data structure that is used in blockchain which encodes the data more efficiently and securely. As shown in Figure 9.6, the hashing starts at the lowest level, that is, leaf level. Hash of each transaction presenting at leaf level are calculated. Similarly, hashing continues to next level, it causes hashes of hashes to ascend until they reach the top root hash. The root hash is called as the Merkle root, it contains all the information about each transaction (Puri et al., 2022, December).

- *Nonce:* The nonce is a random number used in the mining process to find a valid block hash that meets the network's difficulty requirements.
- *Difficulty Target:* It is the largest possible target value which determines the level of complexity required for miners to find a valid block hash during the mining process in a Proof of Work (PoW) consensus algorithm.
- *Transaction:* It contains all transactions, which reflect various activities and data entries. It contains the smart contract (business logic) executions. The transaction involves the exchange or transfer of products, services, or money. Each valid block contains on average 2,000 transactions.
- *Hash (Block Hash):* It is the cryptographic hash of the entire block (Block header and Transaction).
- *Mining Reward:* In some blockchain, the miner is rewarded for mining the block, that is, for adding block to the blockchain. This reward consists of cryptocurrency, tokens, and transaction fees collected from the transaction added in the block (Rani, Pareek, et al., 2023, February).

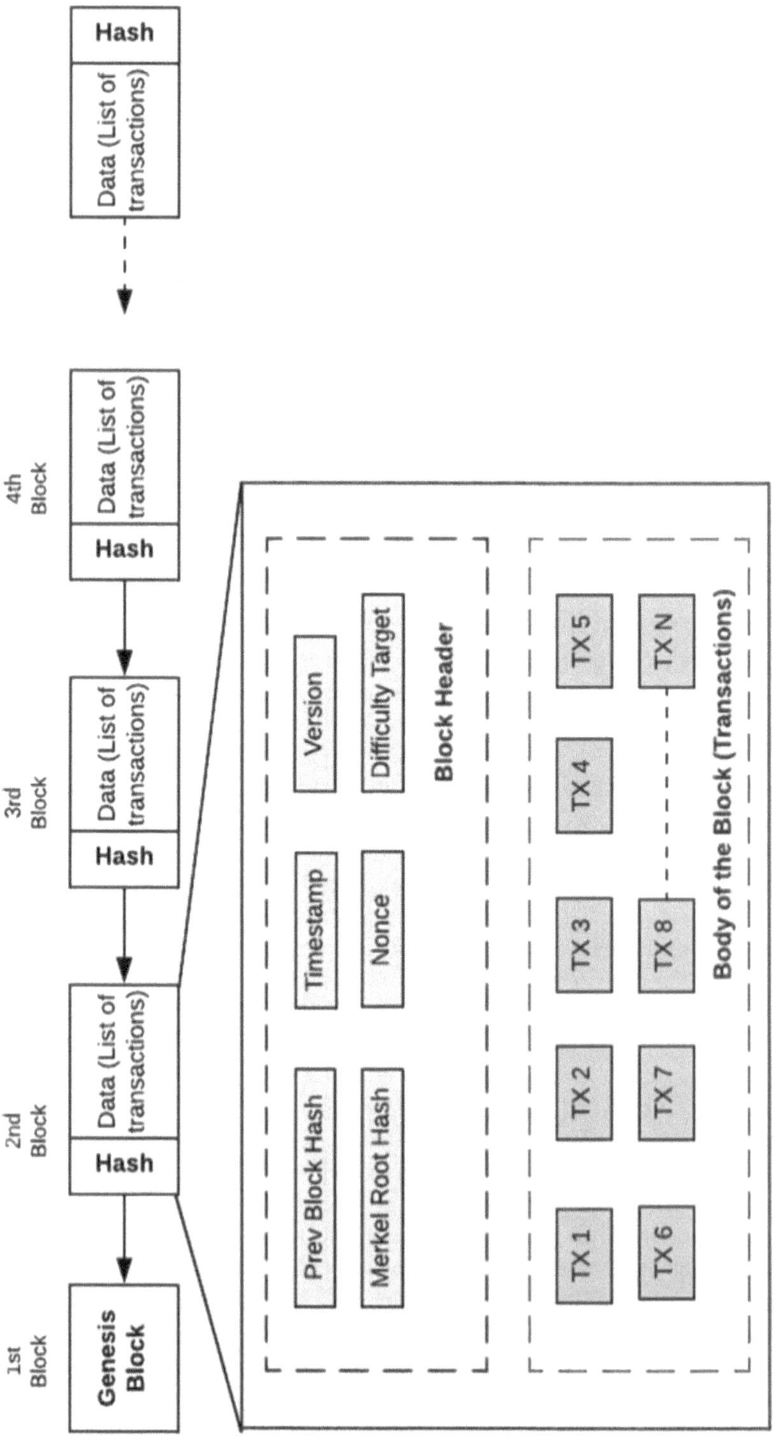

FIGURE 9.5 Structure of a block.

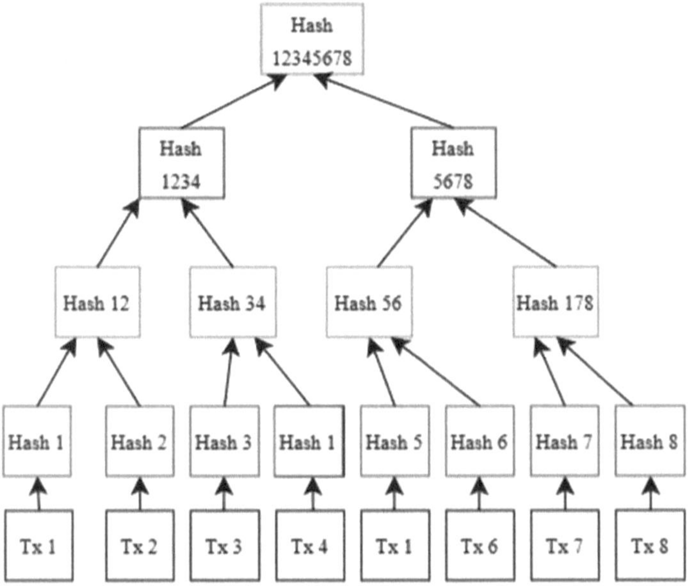

FIGURE 9.6 Merkle tree.

9.2.3.2 Block Generation Process

Transactions created by the user at first enter the system and get authenticated by the smart contract. Then the authenticated transactions are added into pending transaction pool for further processing (Rani, Bhambri, et al., 2023). Miners or validators collect these transactions, check transactions validity, signatures, and whether it adheres to the network rules through the consensus mechanisms. Once the miners or validators receive enough number of valid transactions, they combine them into a block for processing. The cryptographic conundrum will be solved by miners to add the block to the chain. Once the miners find the valid block, they broadcast it into the network and then all the peer nodes in the network verify its validity. The block is added if the greatest possible number of nodes concurs on its authenticity. This block will be linked with the previous block to form the chain of blocks. The miners get reward when they dig a block successfully (Zhang et al., 2023).

9.2.4 Consensus Mechanism

In a typical database, the entire database is managed by a single person or system. It is only their job to upload. But blockchain is self-governing technology where the contribution will involve all the participants or computers around the globe. If one participant does any updation, it is validated by all other participants. By keeping their own records because the database will be open to all participants, everyone will be able to keep track of its current condition. There is no possibility of deception.

To achieve distributed agreement, security, and trust among peers of network, consensus algorithms are used. It allows all participants to achieve integrity and consistency of data and they should reach a common decision on the current state of the network (Hattab & Taha Alyaseen, 2019; Lashkari & Musilek, 2021).

9.2.4.1 Objectives of Consensus Mechanism

- *Unified Agreement:* The public ledger's integrity, as well as the authenticity and accuracy of the data used in the process, is guaranteed by distributed blockchain technology (Rani, Mishra, et al., 2023).
- *Incentive:* A consensus technique promotes good behavior by providing incentives while targeting poor behavior.
- *Fair and Equitable:* Anyone can participate and execute the consensus mechanism in blockchain network.
- *Prevents Double Spending:* Only confirmed and validated transactions should be included in the block, as per the consensus procedure.
- *Fault Tolerant:* Consensus mechanism should work fine even when there were malicious nodes or node failures (Kataria et al., 2023, July).

9.2.4.2 Different Consensus Mechanism

- *Proof of Work*
  It solves problem if everyone has the same copy in network. It is a mechanism which uses the significant man of effort to eliminate the fake uses of computing powers (Bhambri et al., 2023). Here the miners will solve the complex puzzles by defeating each other. These puzzles are very difficult to calculate but very easy to verify the correct solution. It also performs the hashing function on the block. It is mainly used to find the Sybil attacks and solve the double-spending problem. Bitcoin and Litecoin are using PoW (Cachin & Vukolić, 2017; Bentov et al., 2016).
- *Proof of Stake*
  Based on the coins they possess and stake as collateral, PoS chooses validators. Blocks are created and validated by validators, whose chances of selection increase with bigger stakes. For their participation, they receive cryptocurrency (Bhambri & Rani, 2024). PoS uses less energy, promotes users to stake their coins, and has consequences ("slashing") for dishonest behavior. PoS is being used by well-known blockchains like Ethereum 2.0, Cardano, and Tezos to improve scalability, security, and sustainability (Velliangiri & Karthikeyan Karunya, 2020).
- *Proof of Elapsed Time (PoET)*
  Intel introduced this method in 2016. Every participant in the blockchain network must wait for an arbitrary amount of time. The person who completes it will be the leader to propagate the block (Kataria et al., 2022). The proposer's wait time is verified via attestation and the Trusted Execution Environment (TEE) (Rani, Kumar, et al., 2023).
- *Practical Byzantine Fault Tolerance (pBFT)*
  If a rogue node occurs, it is employed in the blockchain to guarantee node agreement. A specified number of nodes must establish consensus before a

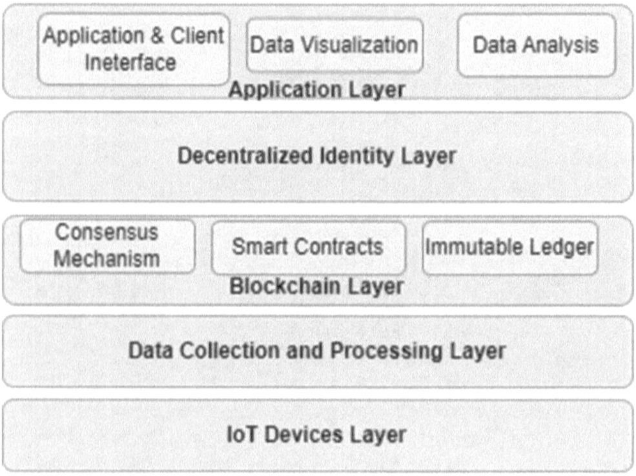

FIGURE 9.7 IoT layer integrated with blockchain.

new block is joined to the chain, which increases the security and dependability of the blockchain. For applications where reliability and consensus are crucial, pBFT is a good choice due to its high throughput and resilience to Byzantine errors.

9.3 BLOCKCHAIN AND IoT

Initially, IoT was implemented using the centralized architecture. The only reason to be concerned about such type of service is that there is security in information sharing, as the IoT nodes do not have any control on information. IoT devices can have the capacity to handle large amount of data due to its architectural design. But they are prone to unexpected and unpredictable types of enemies. But with the use of blockchain in IoT (Ali et al., 2019), the security (Guo & Yu, 2022) level gets increased. IoT has adapted the decentralized architecture of blockchain technology (Alsharari, 2021; Atlam et al., 2018). Figure 9.7 illustrates the IoT layer integrated with blockchain (Rani, Kaur, et al., 2023).

9.3.1 Security Aspects of Integrating Blockchain with IoT

Bringing together blockchain and IoT yields multiple advantages, such as heightened security, transparency, and productivity. Nonetheless, it gives rise to many challenges that can be overcome. In Figure 9.8, we have listed some challenges of integrating blockchain into IoT.

Here we surveyed some papers for addressing different challenges in integrating blockchain with IoT (Alsharari, 2021; Atlam et al., 2020; Guo & Yu, 2022; Kumar & Mallick, 2018; Mohanta et al., 2021; Panarello et al., 2018; Reyna et al., 2018; Sengupta et al., 2020).

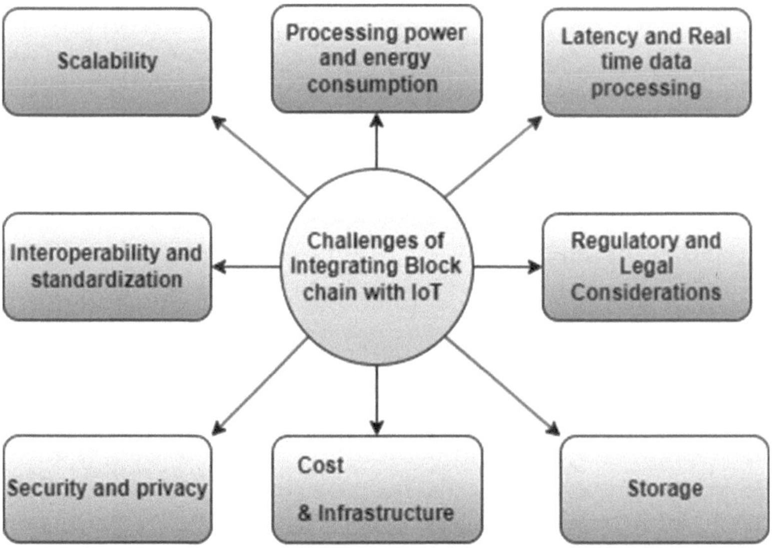

FIGURE 9.8 Different challenges in integrating blockchain.

- *Data Integrity:* Blockchain's immutable feature makes the data to remain in tamper-proof state in IoT. Each data transaction is cryptographically hashed and stored in block. Because of this, it is extremely difficult for malicious nodes to alter or manipulate the data.
- *Secure Authentication and Identity Management:* Blockchains offer a decentralized and secure identity management system. A unique digital identity is provided for each IoT nodes and is stored in blockchain; it ensures secure and authenticated communication between devices. It reduces unauthorized accessing and prevents spoofing or impersonation attacks.
- *Decentralized Trust:* Blockchain enables IoT devices to communicate with one another without the need for any central nodes. The single point of failure is decreased by these decentralization ideas.
- *Secure Communication:* Blockchain creates secure communication for linked devices, boosting data consistency and enabling IoT security through encrypted transactions.
- *Consensus Mechanism:* For the blockchain consensus process to work, all of the network's nodes must concur on the legitimacy of each data transaction. This contributes to the prevention of data tampering and discrepancies resulting in a more dependable and trustworthy IoT environment.
- *Privacy and Consent Management:* By managing access rights, users can exercise greater data ownership. By providing individualized data-sharing control, this satisfies regulatory compliance requirements.
- *Distributed DoS Mitigation:* Since there is no single point of entry or failure, by decentralizing the IoT network on a blockchain, the system becomes more resistant to DDoS attacks.

- *Firmware and Software Updates:* Blockchain is used to distribute updates of the software and firmware of IoT nodes in a secure manner. This confirms that only authorized updates are installed, which reduces the risk of virus or unauthorized modification (Singh & Rani, 2023).
- *Scalability:* IoT nodes will generate a high volume of data which is an issue in blockchain. As in blockchain technology, every node will validate and store transactions. So the blockchain network should be capable to process this data efficiently.
- *Processing Power and Energy Consumption:* In blockchain network, consensus mechanisms such as PoW require high-performance computing resources to solve the complex puzzles and provide result quickly and efficiently. IoT often work with limited resources like low battery life, low memory size, and small size. It will make difficult to implement traditional blockchain in IoT.
- *Latency and Real-Time Data Processing:* IoT in a real-time application requires low latency and high real-time data processing. Distributed ledger and consensus algorithm of blockchain can cause delay as transaction validation and block confirmation are done by all nodes. In IoT settings, maintaining blockchains immutability and security while ensuring timely and efficient data process presents a difficulty.
- *Interoperability and Standardization:* The IoT network consists of diverse types of devices and protocols developed by various manufacturers. It is difficult to achieve compatibility and standardization between these devices and blockchain. For seamless integration and data interchange, it is necessary to develop standard protocols data structure and communication techniques.
- *Legal and Regulatory Considerations:* Issues with law and regulation are brought up by the use of blockchain and IoT. A transparent and immutable blockchain might cause problems with data ownership, permission, and adherence to data privacy laws when sensitive IoT data is stored on it. To maintain compliance and safeguard user privacy, it is crucial to address these legal and regulatory issues.
- *Privacy and Security:* Blockchain will provide security in many aspects in Sybil attacks, DDoS attack. Integrating it with IoT will provide higher level of integrity and security concerns. But IoT nodes are vulnerable to attack and can be easily compromised. Protecting IoT devices from different attacks and providing secure communication is crucial.
- *Cost and Infrastructure:* Blockchain is implemented by large infrastructure investment in hardware, software, and network assets. Deploying IoT nodes with blockchain adds an additional layer of complexity and associated costs. Organization must determine whether implementing blockchain technology for their IoT application is both feasible and cost-effective. For these issues to be resolved and the full potential of blockchain-enabled IoT applications to be realized, it will be essential to build specialized blockchain protocols, consensus mechanisms, and optimization approaches for IoT use cases.

- *Storage:* Local versions of the distributed ledger are maintained locally by blockchain network users. While decentralized storage improves efficiency and new blocks are broadcast, adding IoT data puts a burden on storage. According to a study, a node would require 730 GB annually to exchange 1,000 2-MB photos from participants, creating a storage difficulty.

9.4 CONCLUSION

IoT is a collection of devices which will collect a lot of information. In each and every field, IoT is being used. The devices in the IoT collect sensitive data from the remote environment and send to user for analyzing. One developing technology that can be combined with IoT to enhance security and privacy is blockchain technology. But there is some limitation in embedding IoT with blockchain. In this chapter, we surveyed application and architecture of both the technologies. We even looked into the difficulties involved in combining blockchain and the IoT technology.

REFERENCES

Alam, T. (2018). A reliable communication framework and its use in internet of things (IoT). *International Journal of Scientific Research in Computer Science, Engineering and Information Technology*, 3(5). https://ssrn.com/abstract=3619450

Al-Fuqaha, A., Guizani, M., Mohammadi, M., Aledhari, M., & Ayyash, M. (2015). Internet of things: A survey on enabling technologies, protocols, and applications. *IEEE Communications Surveys and Tutorials*, *17*(4), 2347–2376. https://doi.org/10.1109/COMST.2015.2444095

Ali, M. S., Vecchio, M., Pincheira, M., Dolui, K., Antonelli, F., & Rehmani, M. H. (2019). Applications of blockchains in the internet of things: A comprehensive survey. *IEEE Communications Surveys and Tutorials*, *21*(2), 1676–1717. https://doi.org/10.1109/COMST.2018.2886932

Alsharari, N. (2021). Integrating blockchain technology with internet of things to efficiency. *International Journal of Technology, Innovation and Management (IJTIM)*, *1*(2), 1–13. https://doi.org/10.54489/IJTIM.V1I2.25

Atlam, H. F., Alenezi, A., Alassafi, M. O., & Wills, G. B. (2018). Blockchain with internet of things: Benefits, challenges, and future directions. *International Journal of Intelligent Systems and Applications*, *10*(6), 40–48. https://doi.org/10.5815/IJISA.2018.06.05

Atlam, H. F., Azad, M. A., Alzahrani, A. G., & Wills, G. (2020). A review of blockchain in internet of things and AI. *Big Data and Cognitive Computing*, *4*(4), 28. https://doi.org/10.3390/BDCC4040028

Atzori, L., Iera, A., & Morabito, G. (2010). The internet of things: A survey. *Computer Networks*, *54*(15), 2787–2805. https://doi.org/10.1016/J.COMNET.2010.05.010

Ayoub, W., Ellatif Samhat, A., Mroue, M., Joumaa, H., Nouvel, F., & Prévotet, J. C. (2020). Technology selection for IoT-based smart transportation systems. *Advances in Intelligent Systems and Computing*, *1144*, 19–29. https://doi.org/10.1007/978-981-15-3750-9_2/COVER

Bae, Y., & Chang, H. (2012). Adoption of smart TVs: A Bayesian network approach. *Industrial Management and Data Systems*, *112*(6), 891–910. https://doi.org/10.1108/02635571211238509

Bali, V., Bali, S., Gaur, D., Rani, S., & Kumar, R. (2023). Commercial-off-the shelf vendor selection: A multi-criteria decision-making approach using intuitionistic fuzzy sets and TOPSIS. *Operational Research in Engineering Sciences: Theory and Applications*, *6*(2).

Bentov, I., Gabizon, A., & Mizrahi, A. (2016). Cryptocurrencies without proof of work. *Lecture Notes in Computer Science (Including Subseries Lecture Notes in Artificial Intelligence and Lecture Notes in Bioinformatics), 9604 LNCS*, 142–157. https://doi.org/10.1007/978-3-662-53357-4_10

Bhambri, P., & Rani, S. (2024). Challenges, opportunities, and the future of industrial engineering with IoT and AI. *Integration of AI-Based Manufacturing and Industrial Engineering Systems with the Internet of Things*, 1–18.

Bhambri, P., Rani, S., Balas, V. E., & Elngar, A. A. (Eds.). (2023). *Integration of AI-Based Manufacturing and Industrial Engineering Systems with the Internet of Things*. USA: CRC Press.

Cachin, C., & Vukolić, M. (2017). Blockchain consensus protocols in the wild. *Leibniz International Proceedings in Informatics, LIPIcs, 91*. https://doi.org/10.4230/LIPIcs.DISC.2017.1

Chauhan, M., & Rani, S. (2021). Covid-19: A revolution in the field of education in India. *Learning How to Learn Using Multimedia*, 23–42.

Gubbi, J., Buyya, R., Marusic, S., & Palaniswami, M. (2013). Internet of things (IoT): A vision, architectural elements, and future directions. *Future Generation Computer Systems, 29*(7), 1645–1660. https://doi.org/10.1016/J.FUTURE.2013.01.010

Guo, H., & Yu, X. (2022). A survey on blockchain technology and its security. *Blockchain: Research and Applications, 3*(2), 100067. https://doi.org/10.1016/J.BCRA.2022.100067

Harrison, C., Eckman, B., Hamilton, R., Hartswick, P., Kalagnanam, J., Paraszczak, J., & Williams, P. (2010). Foundations for smarter cities. *IBM Journal of Research and Development, 54*(4), 1. https://doi.org/10.1147/jrd.2010.2048257

Hattab, S., & Taha Alyaseen, I. F. (2019). Consensus algorithms blockchain: A comparative study. *International Journal on Perceptive and Cognitive Computing, 5*(2), 66–71. https://doi.org/10.31436/IJPCC.V5I2.103

Kataria, A., Agrawal, D., Rani, S., Karar, V., & Chauhan, M. (2022). Prediction of blood screening parameters for preliminary analysis using neural networks. In *Predictive Modeling in Biomedical Data Mining and Analysis* (pp. 157–169). New York, NY: Academic Press.

Kataria, A., Puri, V., Pareek, P. K., & Rani, S. (2023, July). Human activity classification using G-XGB. In *2023 International Conference on Data Science and Network Security (ICDSNS)* (pp. 1–5). Tiptur: IEEE.

Kim, T. H., Ramos, C., & Mohammed, S. (2017). Smart city and IoT. In *Future Generation Computer Systems* (Vol. 76, pp. 159–162). Elsevier B.V. https://doi.org/10.1016/j.future.2017.03.034

Kortuem, G., Kawsar, F., Fitton, D., & Sundramoorthy, V. (2010). *Internet of Things Track Smart Objects as Building Blocks for the Internet of Things*. www.computer.org/internet/

Kumar, N. M., & Mallick, P. K. (2018). Blockchain technology for security issues and challenges in IoT. *Procedia Computer Science, 132*, 1815–1823. https://doi.org/10.1016/J.PROCS.2018.05.140

Kumar, P., Banerjee, K., Singhal, N., Kumar, A., Rani, S., Kumar, R., & Lavinia, C. A. (2022). Verifiable, secure mobile agent migration in healthcare systems using a polynomial-based threshold secret sharing scheme with a Blowfish algorithm. *Sensors, 22*(22), 8620.

Laouiti, A., Qayyum, A., & Mohamad Saad, M. N. (Eds.). (2020). *Vehicular Ad-hoc Networks for Smart Cities* (Vol. 1144). https://doi.org/10.1007/978-981-15-3750-9

Lashkari, B., & Musilek, P. (2021). A comprehensive review of blockchain consensus mechanisms. *IEEE Access, 9*, 43620–43652. https://doi.org/10.1109/ACCESS.2021.3065880

Meunier, S. (2018). Blockchain 101: What is blockchain and how does this revolutionary technology work? What is blockchain and how does this revolutionary technology work? In *Transforming Climate Finance and Green Investment with Blockchains* (pp. 23–34). Elsevier. https://doi.org/10.1016/B978-0-12-814447-3.00003-3

Miorandi, D., Sicari, S., De Pellegrini, F., & Chlamtac, I. (2012). Internet of things: Vision, applications and research challenges. *Ad Hoc Networks*, *10*(7), 1497–1516. https://doi.org/10.1016/J.ADHOC.2012.02.016

Mohanta, B. K., Jena, D., Ramasubbareddy, S., Daneshmand, M., & Gandomi, A. H. (2021). Addressing security and privacy issues of IoT using blockchain technology. *IEEE Internet of Things Journal*, *8*(2), 881–888. https://doi.org/10.1109/JIOT.2020.3008906

Pal, A., Mukherjee, A., & Dey, S. (2016). *Future of Healthcare—Sensor Data-Driven Prognosis* (pp. 93–109). https://doi.org/10.1007/978-3-319-42141-4_9

Panarello, A., Tapas, N., Merlino, G., Longo, F., & Puliafito, A. (2018). Blockchain and IoT integration: A systematic survey. *Sensors (Switzerland)*, *18*(8). https://doi.org/10.3390/S18082575

Pannayagol, B. B., & Deshpande, S. (2023). Security in internet of things: An overview. *Proceedings—IEEE International Conference on Device Intelligence, Computing and Communication Technologies, DICCT 2023*, 243–248. https://doi.org/10.1109/DICCT56244.2023.10110070

Puri, V., Kataria, A., Solanki, V. K., & Rani, S. (2022, December). AI-based botnet attack classification and detection in IoT devices. In *2022 IEEE International Conference on Machine Learning and Applied Network Technologies (ICMLANT)* (pp. 1–5). Salvador: IEEE.

Rani, S., Bhambri, P., & Kataria, A. (2023). Integration of IoT, big data, and cloud computing technologies. In *Big Data, Cloud Computing and IoT: Tools and Applications*. USA: CRC Press.

Rani, S., Kataria, A., & Chauhan, M. (2022). Cyber security techniques, architectures, and design. In *Holistic Approach to Quantum Cryptography in Cyber Security* (pp. 41–66). USA: CRC Press.

Rani, S., Kataria, A., Kumar, S., & Tiwari, P. (2023). Federated learning for secure IoMT-applications in smart healthcare systems: A comprehensive review. *Knowledge-Based Systems*, 110658.

Rani, S., Kaur, J., & Bhambri, P. (2023). Technology and gender violence: Victimization model, consequences and measures. In *Communication Technology and Gender Violence* (pp. 1–19). Cham: Springer International Publishing.

Rani, S., Kumar, S., Kataria, A., & Min, H. (2023). SmartHealth: An intelligent framework to secure IoMT service applications using machine learning. *ICT Express*, *10*(2), 420–425.

Rani, S., Mishra, A. K., Kataria, A., Mallik, S., & Qin, H. (2023). Machine learning-based optimal crop selection system in smart agriculture. *Scientific Reports*, *13*(1), 15997.

Rani, S., Pareek, P. K., Kaur, J., Chauhan, M., & Bhambri, P. (2023, February). Quantum machine learning in healthcare: Developments and challenges. In *2023 IEEE International Conference on Integrated Circuits and Communication Systems (ICICACS)* (pp. 1–7). Raichur: IEEE.

Reyna, A., Martín, C., Chen, J., Soler, E., & Díaz, M. (2018). On blockchain and its integration with IoT. Challenges and opportunities. *Future Generation Computer Systems*, *88*, 173–190. https://doi.org/10.1016/J.FUTURE.2018.05.046

Sengupta, J., Ruj, S., & Das Bit, S. (2020). A comprehensive survey on attacks, security issues and blockchain solutions for IoT and IIoT. *Journal of Network and Computer Applications*, *149*. https://doi.org/10.1016/J.JNCA.2019.102481

Sharma, N., Shamkuwar, M., & Singh, I. (2019). The history, present and future with IoT. *Intelligent Systems Reference Library*, *154*, 27–51. Springer Science and Business Media Deutschland GmbH. https://doi.org/10.1007/978-3-030-04203-5_3/COVER

Singh, H., & Rani, S. (2023). Flex sensor integrated smart strap to verify correct wearing of the face mask. *IEEE Sensors Journal*, *24*(2), 2020–2027.

Sobin, C. C. (2020). A survey on architecture, protocols and challenges in IoT. *Wireless Personal Communications*, *112*(3), 1383–1429. Springer. https://doi.org/10.1007/s11277-020-07108-5

Sudevan, S., Barwani, B., Al Maani, E., Rani, S., & Sivaraman, A. K. (2021). Impact of blended learning during Covid-19 in Sultanate of Oman. *Annals of the Romanian Society for Cell Biology*, 14978–14987.

Swan, M. (2015). Blockchain thinking: The brain as a decentralized autonomous corporation. *IEEE Technology and Society Magazine*, *34*(4), 41–52. https://doi.org/10.1109/mts.2015.2494358

Tanwar, R., Chhabra, Y., Rattan, P., & Rani, S. (2022, September). Blockchain in IoT networks for precision agriculture. In *International Conference on Innovative Computing and Communications: Proceedings of ICICC 2022* (Vol. 2, pp. 137–147). Singapore: Springer Nature Singapore.

Velliangiri, S., & Karthikeyan Karunya, P. (2020). Blockchain technology: Challenges and security issues in consensus algorithm. *2020 International Conference on Computer Communication and Informatics, ICCCI 2020*. https://doi.org/10.1109/ICCCI48352.2020.9104132

Zhang, C., Ma, Z., Liu, L., & Liu, Y. (2023). Modeling & analysis of block generation process of the mining pool in blockchain system. *Peer-to-Peer Networking and Applications*, *16*(2), 475–487. https://doi.org/10.1007/S12083-022-01359-8/TABLES/6

Zheng, Z., Xie, S., Dai, H. N., Chen, X., & Wang, H. (2018). Blockchain challenges and opportunities: A survey. *International Journal of Web and Grid Services*, *14*(4), 352–375. https://doi.org/10.1504/IJWGS.2018.095647

10 Exploring Integration of the Blockchain and Internet of Things (IoT) Computing for Secure and Decentralized Data Management

Tarun Kumar Vashishth, Vikas Sharma, Kewal Krishan Sharma, Bhupendra Kumar, Sachin Chaudhary, and Rajneesh Panwar

10.1 INTRODUCTION

The rapid advancements in technology have ushered in a new era of connectivity and data proliferation. Both techniques demonstrated their transformative potential in various domains, and their integration holds the promise of revolutionizing data management, security, and decentralization. Blockchain, originally introduced as the underlying technology for cryptocurrencies like Bitcoin, has transcended its initial application to become a decentralized ledger with immense capabilities. It offers unprecedented levels of data security, immutability, and transparency, making it a formidable solution for mitigating data breaches and unauthorized access. On the other hand, the Internet of Things (IoT) encompasses a network of interconnected devices capable of gathering and exchanging data. These could generate vast amounts of data, contributing to a hyper-connected ecosystem. The convergence of blockchain and IoT presents a compelling synergy that addresses critical challenges in data management. The integration promises enhanced data integrity, secure data-sharing, decentralized governance, and improved privacy for the ever-expanding IoT landscape. This chapter delves into the profound implications of this integration, exploring the potential to establish secure and decentralized data management systems that empower individuals and organizations alike.

 DOI: 10.1201/9781003460367-10

Objectives of the Chapter:

- The primary objective of this chapter is to provide a comprehensive exploration of the integration of blockchain and IoT computing for secure and decentralized data management. By delving into the intricate interplay between these two transformative technologies, we aim at the following:
- Examine the foundational concepts of blockchain and IoT and their relevance to contemporary data management challenges.
- Highlight the inherent security and privacy challenges within the IoT ecosystem and how blockchain can offer robust solutions.
- Explore the concept of decentralized data governance enabled by smart contracts, ensuring transparency, traceability, and tamper-proof data records.
- Investigate real-world case studies that demonstrate successful implementations of blockchain–IoT fusion in various industries.
- Discuss potential challenges, limitations, and opportunities that arise from this integration.
- Anticipate future trends, research directions, and the role of standardization in shaping the landscape of blockchain–IoT data management.

10.1.1 Introduction to Blockchain and IoT Integration

The convergence of blockchain technology and IoT has ignited a paradigm shift in how data is managed, secured, and exchanged across interconnected devices and systems. This section delves into the foundational concepts of integrating blockchain and IoT, highlighting the driving forces behind this convergence and setting the stage for exploring the multifaceted dimensions of secure and decentralized data management.

Understanding Blockchain's Role: Blockchain, at its core, is a distributed and immutable ledger that offers transparency, security, and trust in recording transactions. Originally conceptualized for cryptocurrencies, its application has transcended into domains where data integrity and tamper-proof records are paramount. By employing cryptographic techniques and decentralized consensus mechanisms, blockchain establishes a system that resists unauthorized modifications, fostering an environment of verifiable and auditable data.

Exploring the Potential of IoT: The Internet of Things (IoT) represents a network of interconnected devices that collect, share, and transmit data through the internet. These devices, ranging from everyday objects to sophisticated sensors, form an intricate ecosystem capable of real-time data exchange and automation. The proliferation of IoT devices has revolutionized industries by enabling real-time monitoring, predictive analytics, and informed decision-making based on data insights.

Fusion for Data Transformation: The symbiotic relationship between IoT and blockchain addresses crucial challenges in data management. By combining blockchain's attributes of immutability, transparency, and decentralized consensus with IoT's vast data generation capacity, a new era of secure and decentralized data management emerges. This fusion aims to establish data integrity, ownership, and accessibility while minimizing vulnerabilities inherent in centralized systems (Figure 10.1).

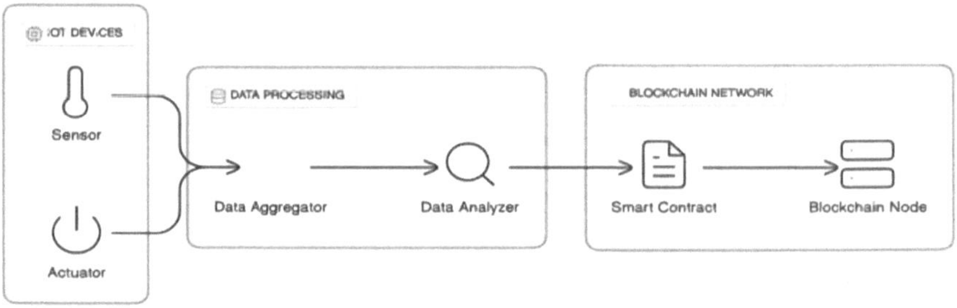

FIGURE 10.1 Integration of blockchain technology and IoT.

10.1.2 Significance of Secure and Decentralized Data Management

The significance of secure and decentralized data management lies at the core of the integration between blockchain and Internet of Things (IoT) computing. This section delves into the compelling reasons why these technologies are being fused to address the critical challenges surrounding data security, integrity, privacy, and ownership in the modern digital landscape.

Ensuring Data Integrity: In an era where data fuels decision-making and operations across industries, maintaining data integrity is paramount. An effective way to counteract data breaches, illegal changes, and tampering is through the integration of blockchain with IoT. Because of blockchain's immutability, data cannot be changed after it has been recorded, creating a reliable history of transactions and occurrences.

Minimizing Centralized Vulnerabilities: Decentralized blockchain–IoT integration distributes data across a network of nodes, reducing the risk of a single breach compromising the entire system. This decentralization also empowers stakeholders by giving them greater control over their data, mitigating concerns about a single entity monopolizing or mishandling sensitive information.

Enhancing Data Privacy: In an age of increasing data privacy concerns, the amalgamation of blockchain and IoT presents a robust approach to data protection. Through cryptographic techniques and private key ownership, blockchain ensures that data remains confidential and accessible only to authorized parties. This is particularly crucial in scenarios where personal, sensitive, or confidential data is involved.

Empowering Data Ownership: The conventional model of data ownership often tilts in favor of centralized entities, leaving individuals and smaller participants with limited control over their data. Blockchain's transparent and decentralized nature allows stakeholders the ability to verify, share, and even monetize their information on their terms.

Transparency and Accountability: The transparency inherent in blockchain enables traceability and accountability across the life cycle of data. In the context of IoT, this means that the origin, journey, and transformations of data can be tracked and audited. This level of transparency fosters trust among stakeholders and streamlines compliance with regulatory requirements.

Unlocking New Possibilities: The secure and decentralized data management facilitated by blockchain–IoT integration paves the way for innovative applications and business models. From supply chain traceability and real-time asset monitoring to smart contracts automating data-driven agreements, the synergy between these technologies fosters a fertile ground for novel solutions. The integration of blockchain and IoT addresses the contemporary challenges of data management with a focus on security, transparency, and autonomy. As industries evolve in the digital age, the significance of secure and decentralized data management becomes undeniable, making the exploration of this fusion not only relevant but imperative for shaping the future of data-driven ecosystems.

10.2 LITERATURE REVIEW

Xiong et al. [1] present a case study introducing a learning-assisted resource allocation approach for intelligent data management. The study's numerical outcomes demonstrate the superior performance of the proposed method when compared to baseline solutions. Shafagh et al. [2] introduce a blockchain-based IoT design featuring distributed access control and data management. Their design deviates from the conventional trust model, allowing users to possess data ownership and enabling secure data-sharing. The proposal establishes a secure and robust access control management by leveraging the blockchain as an auditable and distributed layer for access control within the storage layer. Novo [3] presents a novel architecture introducing distributed access control for IoT through blockchain technology, encompassing role and permission arbitration. The architecture is substantiated by a proof-of-concept implementation and assessment within practical IoT contexts. Pinno et al. [4] introduce a blockchain-based architecture for IoT access authorizations, designed to align with IoT requirements. Their user-transparent, decentralized, scalable, and fault-tolerant architecture is crafted to seamlessly accommodate various access control models used in the IoT ecosystem.

Luong et al. [5] demonstrate experimental outcomes that substantiate the advantages of employing deep learning to derive optimal auctions for mobile blockchains, resulting in enhanced revenue. Datiri et al. [6] present a three-tiered framework merging clustering, edge computing, and blockchain to enhance the data management mechanisms within the blockchain architecture and data structure. This framework aims to establish a decentralized, secure approach for resource optimization and data management in IoT systems, featuring permissioned authentication, smart contract generation, and immutable block transactions across nodes, without being confined to a single node.

Okegbile et al. [7] explore a collaborative data-sharing approach, wherein multiple data providers and users collaborate through blockchain and cloud-edge computing techniques. They model the spatial distribution of providers and users as independent homogeneous Poisson point processes and describe transaction generation rates using independent Bernoulli processes. Yang et al. [8] introduce ORAC, an optimized encrypted access control system with partial privacy and reliability, constructed upon functional encryption to ensure data privacy and utilization in

VANET. The system segregates data access into offline and online phases, allowing substantial decryption computation during the offline phase and efficient real decryption by vehicles. Naghib et al. [9] conduct an in-depth comparison of mechanisms within each category, culminating in a discussion of the developmental challenges and unresolved aspects of blockchain data management (BDM) in IoT. Tian [10] introduces a dynamically distributed internet database architecture that facilitates intricate decision-making regarding fragmentation, distribution, and duplication. Shen et al. [11] explore blockchain solutions to tackle this challenge, presenting research issues and potential solutions for blockchain-based data management (DM) toward 6G from these three perspectives. Peng et al. [12] formulate the mathematical description and modeling method for the lattice matching decision problem, while also introducing an artificial neural network designed to discover the problem's optimal solution.

10.3 FUNDAMENTALS OF BLOCKCHAIN AND IoT

10.3.1 Overview of Blockchain Technology

Blockchain technology, a revolutionary innovation underpinning cryptocurrencies like Bitcoin, has transcended its origins to revolutionize various industries. This section provides a comprehensive overview of the fundamental aspects of blockchain technology, unraveling its core principles and mechanisms:

Decentralization: At the heart of blockchain lies the concept of decentralization, where a network of nodes collaboratively validates and records transactions. This eliminates the need for a central authority, distributing control and decision-making among participants.

Immutable Ledger: The blockchain ledger is immutable, meaning once data is recorded on a block, it cannot be altered or deleted. This feature is achieved through cryptographic hashing and consensus mechanisms.

Transparency: Transparency is a hallmark of blockchain technology. All participants in the network have access to the entire transaction history, promoting trust and accountability.

Consensus Mechanisms: Blockchain networks employ various consensus mechanisms to ensure agreement on the validity of transactions. These mechanisms include Proof of Work (PoW), Proof of Stake (PoS), and others.

Smart Contracts: Smart contracts are self-executing agreements with predefined conditions. They automate and facilitate the execution of transactions based on these conditions, reducing the need for intermediaries.

Cryptographic Security: Blockchain relies on cryptographic techniques to secure data and ensure confidentiality, integrity, and authentication of transactions.

Public versus Private Blockchains: Blockchains can be public, allowing anyone to participate, or private, where access is restricted to authorized entities. The choice depends on the use case and desired level of control.

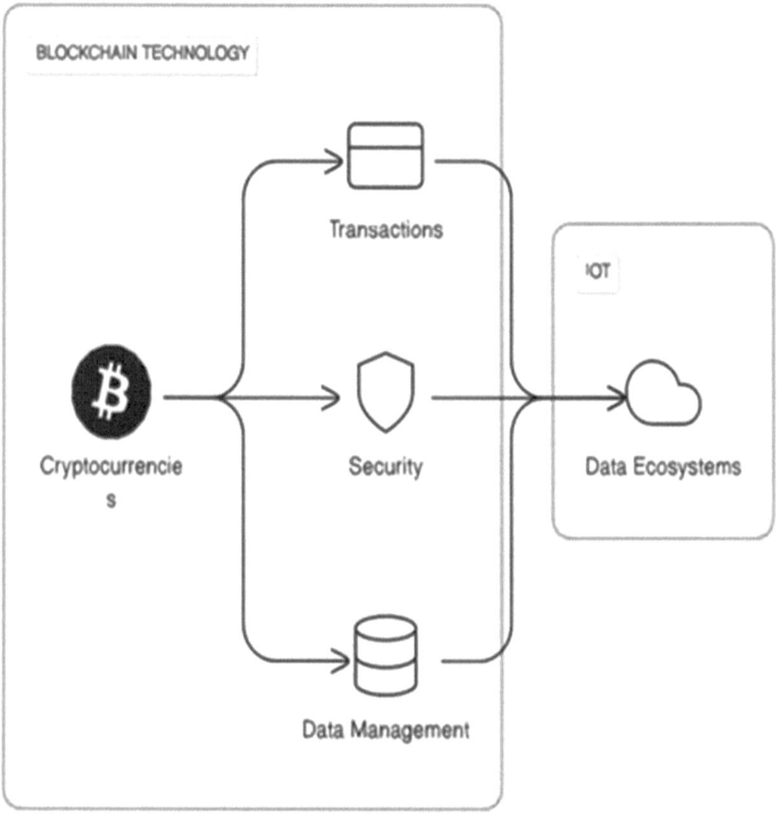

FIGURE 10.2 Cloud structure of blockchain technology.

Understanding the core components of blockchain technology lays the groundwork for appreciating its role in revolutionizing data management, security, and transactions across industries (Figure 10.2). This foundational understanding of blockchain technology will illuminate how it synergizes with IoT to create secure, decentralized, and transparent data ecosystems.

10.3.2 Introduction to Internet of Things (IoT) Computing

The Internet of Things (IoT) has ushered in a new era of connectivity, where devices seamlessly communicate, collect, and exchange data, revolutionizing industries and enhancing the way we interact with our environment (Figure 10.3).

This section provides an in-depth introduction to IoT computing, shedding light on its essential components and transformative capabilities:

Device Diversity: IoT encompasses a diverse array of devices, from everyday objects like thermostats and wearable to industrial machinery and smart

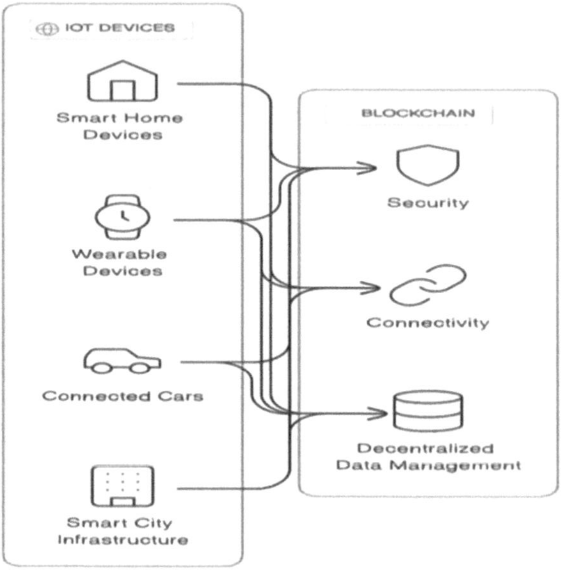

FIGURE 10.3 IoT in daily life.

vehicles. These devices are embedded with sensors, processors, and communication modules, enabling data collection and transmission.

Data Generation and Collection: IoT devices generate a continuous stream of data as they interact with their surroundings. This data is collected in real time and can be harnessed for various purposes, such as monitoring, analysis, and automation.

Interconnectivity: IoT devices communicate with each other and with central systems through the internet or other communication protocols. This interconnectivity enables seamless data exchange and remote control.

Automation and Intelligence: IoT empowers automation by leveraging data insights. Devices can autonomously respond to data triggers, making real-time decisions and executing actions without human intervention.

Real-Time Insights: By providing real-time data, IoT enables informed decision-making. This is particularly valuable in scenarios where timely actions are crucial, such as predictive maintenance in industrial settings or health monitoring in healthcare.

Ubiquitous Connectivity: IoT computing extends beyond traditional devices to create a pervasive network where virtually anything can be connected and communicate, from streetlights to agricultural sensors.

Challenges and Opportunities: While IoT computing offers immense opportunities, challenges include data privacy, security, interoperability, and scalability.

The rapid proliferation of IoT devices and their integration into daily life underscores the transformative potential of this technology. Understanding IoT's fundamental principles is essential for grasping how it synergizes with blockchain to create secure, decentralized, and data-driven ecosystems that redefine industries and user experiences. As we navigate through the integration of blockchain and IoT, this foundational knowledge of IoT computing will illuminate how these technologies harmonize to usher in a new era of data management, security, and connectivity.

10.3.3 Synergies and Challenges in Integrating Blockchain and IoT

The convergence of blockchain and Internet of Things (IoT) offers a transformative synergy that promises enhanced data security, transparency, and decentralized control. However, this integration also presents its share of challenges. This section delves into the synergies and challenges that arise when combining these two pioneering technologies:

Synergies

Data Integrity and Security: Blockchain's immutability is particularly crucial in scenarios where the accuracy of data is paramount, such as supply chain tracking or critical infrastructure monitoring.

Decentralization and Trust: Both blockchain and IoT share a common goal of decentralization. By eliminating intermediaries and enabling direct device-to-device communication, this fusion enhances trust and reduces the need for centralized authorities.

Transparency and Auditability: Blockchain's transparent and open nature aligns with IoT's need for traceability. Participants in the network can verify the origin and journey of data, enhancing accountability and streamlining compliance.

Smart Contracts for Automation: Smart contracts, a feature of blockchain, can automate and execute actions based on predefined conditions. In an IoT context, this means that devices can interact and trigger actions autonomously without human intervention.

Challenges

Scalability: The massive data generated by IoT devices poses scalability challenges for blockchain networks. Ensuring that the blockchain can handle the volume of transactions and data is essential for a successful integration.

Latency: IoT applications often require real-time data processing and decision-making. The time required for consensus in blockchain networks can introduce latency that might not be compatible with IoT's real-time demands.

Energy Efficiency: Many IoT devices operate with energy constraints. The resource-intensive nature of some blockchain operations could clash with IoT's goal of energy-efficient operation.

Interoperability: IoT devices come from various manufacturers and may use different communication protocols. Ensuring interoperability between diverse devices and blockchain networks can be complex.

Cost and Complexity: Integrating blockchain with IoT adds complexity to the ecosystem. The costs associated with implementing and maintaining these technologies can be a barrier, particularly for smaller players.

Understanding the synergies and challenges that arise from integrating blockchain and IoT is vital for designing effective solutions that harness the benefits while mitigating the obstacles. As we navigate through the exploration of this integration, these insights will guide us in comprehending the intricate dance between blockchain and IoT and the opportunities it unveils [13–19].

10.4 DECENTRALIZED DATA GOVERNANCE AND OWNERSHIP

10.4.1 Smart Contracts for Data Access and Sharing

In the landscape of integrating blockchain and IoT, the establishment of decentralized data governance and ownership takes center stage. This section delves into the transformative role of smart contracts in shaping data access and sharing within this fusion:

Automated Agreements: Smart contracts, self-executing code residing on the blockchain, empower seamless automation of data access agreements. In the context of IoT-generated data, these contracts can facilitate predefined conditions for data-sharing and access. For example, a smart contract could enable a weather sensor to automatically share its data with a relevant agricultural application based on certain weather conditions [20–23].

Data Monetization: Smart contracts offer a gateway for IoT devices to monetize their data. By creating contracts that enable authorized access to specific data streams, IoT devices can be compensated with cryptocurrencies or tokens for sharing valuable data. This concept transforms IoT devices into data vendors, fostering a marketplace for data exchange.

Enhanced Security: The utilization of smart contracts ensures that data-sharing adheres to pre-established terms. Data access is automatically granted once the smart contract conditions are met, enhancing security by minimizing manual intervention and reducing potential vulnerabilities.

Decentralized Control: Smart contracts decentralize the control over data access, shifting it from a centralized authority to a trustless, code-driven process. This transformation aligns with the foundational principles of blockchain technology, where intermediaries are eliminated in favor of peer-to-peer interactions [24–26].

Immutable Record: As smart contract executions are recorded on the blockchain, an immutable record of data-sharing agreements is created. This transparent and tamper-proof record enhances accountability and transparency, building trust among participants.

In essence, smart contracts act as digital enforcers of data-sharing agreements, streamlining the process while upholding transparency and security. As we navigate further into the interplay between blockchain and IoT, this understanding of smart contracts will illuminate how they shape the landscape of data governance and ownership in a decentralized ecosystem [27–29].

10.4.2 Ownership and Control of IoT Data

In the realm of integrating blockchain and the Internet of Things (IoT), the concept of ownership and control over data emerges as a pivotal consideration. This section delves into the implications of this integration on the fundamental aspects of data ownership and control:

Decentralized Ownership: The fusion of blockchain and IoT introduces a paradigm shift in data ownership. Unlike centralized models where data is controlled by a single entity, blockchain's decentralized nature empowers individual users and devices with ownership and control over their data. Each participant becomes a sovereign owner of their generated data [30].

Permissioned Access: Blockchain's transparency, coupled with its ability to implement fine-grained access controls through smart contracts, empowers users to grant permissioned access to their data. This dynamic model ensures that data is shared only with authorized entities or individuals, preserving privacy and security.

Auditable History: Blockchain's inherent feature of immutability ensures that every data interaction is recorded on the distributed ledger. This audit trail provides a transparent and tamper-proof history of data access, changes, and transfers, fostering accountability and trust.

Immutable Records of Ownership: Once data ownership is established and recorded on the blockchain, it becomes an immutable record. This prevents disputes and ambiguity over data ownership, especially in scenarios where data is shared or monetized among various stakeholders.

Empowerment of Data Producers: The integration of blockchain and IoT shifts the power dynamics by giving data producers, such as IoT devices or individuals, greater control over their data. This is particularly significant in cases where data monetization or sharing is involved.

Disintermediation: Blockchain and IoT remove the requirements of traditionally control data access. This disintermediation not only enhances data security but also simplifies data-sharing processes.

10.4.3 Data Marketplace and Monetization in IoT–Blockchain Integration

The synergy between blockchain and the Internet of Things (IoT) not only redefines data governance but also paves the way for innovative data marketplaces and monetization strategies (Figure 10.4). This section delves into the emergence of data marketplaces within this integration:

Data Trading Platforms: Blockchain's tamper-proof nature ensures data integrity, while its transparency fosters trust among participants.

Automated Transactions: Smart contracts play a pivotal role in data marketplaces, automating transactions and ensuring that data-sharing adheres to predefined terms. Once certain conditions are met, smart contracts

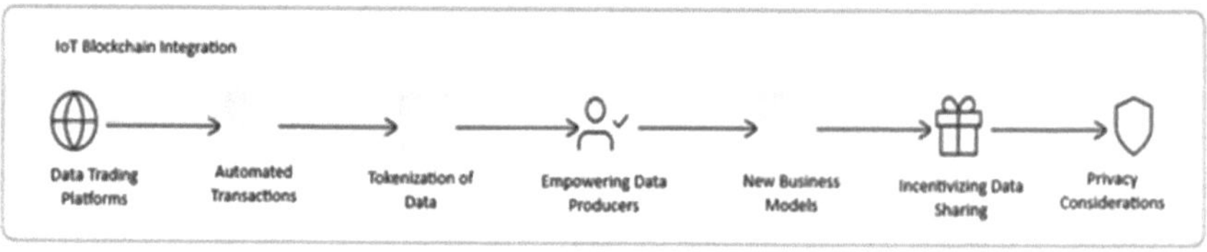

FIGURE 10.4 Data marketplace and monetization in IoT–blockchain integration.

automatically facilitate data exchange, eliminating intermediaries and expediting the process.

Tokenization of Data: The concept of tokenization transforms data into digital assets that can be traded or exchanged. IoT-generated data can be tokenized, enabling fractional ownership and granular transactions. This token-based system also facilitates micropayments, enabling even small data contributions to be monetized.

Empowering Data Producers: Data marketplaces shift the power dynamics by allowing data producers, such as IoT devices or individuals, to take ownership of their data's value. This enables them to directly engage in the data economy and be compensated for their contributions.

New Business Models: Data marketplaces open avenues for new business models. IoT devices can generate revenue by providing valuable data to interested parties. Additionally, data consumers can access specific data streams without the need for complex negotiation processes.

Incentivizing Data-Sharing: Blockchain's token-based mechanisms can incentivize IoT users to contribute data to the marketplace. This creates a self-sustaining ecosystem where data-sharing benefits all participants and enriches the overall data pool.

Privacy Considerations: While data marketplaces offer monetization opportunities, privacy remains paramount. Blockchain's cryptographic techniques and consent-driven data-sharing can ensure that sensitive information is protected.

10.5 SECURE DATA MANAGEMENT WITH BLOCKCHAIN AND IoT

10.5.1 Data Integrity and Immutability through Blockchain

The convergence of blockchain and the Internet of Things (IoT) introduces a powerful synergy for secure data management.

This section explores the foundational aspects of data integrity and immutability enabled by blockchain (Figure 10.5):

Immutable Ledger: Blockchain's distributed ledger ensures data immutability by recording transactions in a tamper-proof manner. Each data entry, or "block," is cryptographically linked to the previous one, creating an unbreakable chain of information.

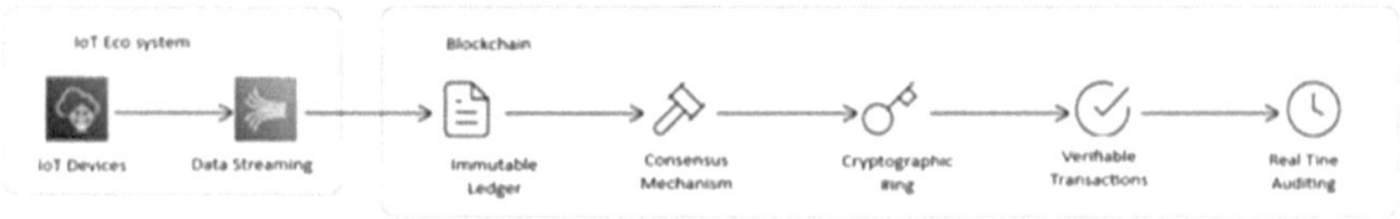

FIGURE 10.5 Immutability enabled by blockchain.

Ensuring Data Integrity: In an IoT ecosystem, where data streams from various devices, ensuring the integrity of data is crucial. Blockchain's consensus mechanisms and cryptographic hashing maintain the accuracy of data by preventing unauthorized alterations.

Verifiable Transactions: Participants in the network can verify the authenticity of data by tracing its origin and confirming its journey through the blockchain. This transparency builds trust and confidence in the accuracy of shared data.

Real-Time Auditing: Blockchain enables real-time auditing of data as it flows through the IoT ecosystem. This feature is especially valuable for compliance, regulatory reporting, and quality control purposes.

10.5.2 IoT Devices as Data Sources and Sensors

IoT devices act as valuable data sources and sensors, generating a constant stream of information. This section highlights the significance of these devices in the context of secure data management:

Data Generation: IoT devices collect data from the physical world, ranging from environmental parameters to user interactions. This data forms the basis for informed decision-making and automation.

Data Variety: IoT devices produce diverse data types, including structured and unstructured data, images, videos, and sensor readings. This diversity enhances the richness and depth of the available data.

Decentralized Data Points: The distributed nature of IoT devices decentralizes data points, reducing the risk of a single point of failure or manipulation.

10.5.3 Ensuring Data Authenticity and Trustworthiness

Integrating blockchain with IoT bolsters data authenticity and trustworthiness, addressing concerns related to data manipulation and unauthorized access:

Digital Signatures: IoT devices can cryptographically sign data, creating a digital fingerprint unique to each device. Blockchain verifies the authenticity of these signatures, ensuring data integrity.

Data Provenance: Blockchain records the origin and journey of data, providing an auditable trail. This provenance enhances transparency and accountability, ensuring that data can be trusted and traced back to its source.

Secure Identity Management: Blockchain's decentralized identity management ensures that only authorized devices can contribute or access data. This prevents unauthorized data injection and manipulation.

In essence, the secure data management enabled by the integration of blockchain and IoT ensures data authenticity, immutability, and trustworthiness. These aspects collectively create a foundation for robust data-driven decision-making, automation, and innovative applications in diverse industries.

10.6 PRIVACY AND CONSENT MANAGEMENT

10.6.1 Privacy Challenges in IoT Data Collection

The convergence of blockchain and the Internet of Things (IoT) ushers in transformative opportunities, yet it also presents intricate privacy challenges related to data collection. This section delves into the multifaceted privacy challenges arising from the proliferation of IoT-generated data:

Data Proliferation: The widespread deployment of IoT devices generates an unprecedented volume of data. This deluge of information increases the risk of unintentional data exposure, potentially leading to privacy breaches.

Granularity of Data: IoT devices intricately capture user behaviors, activities, and interactions in real time. This fine-grained data collection raises concerns about the potential inference of sensitive details and unauthorized profiling.

Intrusive Monitoring: IoT devices often monitor users in their personal spaces, such as homes or wearable devices. This constant monitoring raises privacy concerns, as users may feel uncomfortable with the level of surveillance.

Data-Sharing with Third Parties: In IoT ecosystems, data-sharing is not limited to device manufacturers and users. The interconnected nature of IoT raises the possibility of data being shared with third parties, amplifying the risk of unintended data exposure.

Security Vulnerabilities: IoT devices are susceptible to security vulnerabilities and breaches, which can expose sensitive data to malicious actors. These breaches can lead to severe privacy implications.

As we navigate through the landscape of IoT data collection within the realm of blockchain integration, addressing these privacy challenges becomes a paramount consideration to ensure that the benefits of this fusion are realized without compromising individual privacy rights.

10.6.2 Consent Mechanisms through Smart Contracts

As the integration of blockchain and the Internet of Things (IoT) evolves, innovative solutions are required to address privacy concerns. This section explores how smart contracts can serve as potent tools for managing user consent within this dynamic landscape:

Automated Consent Execution: Smart contracts, self-executing code residing on the blockchain, offer an automated approach to executing user-defined consent agreements. Users can specify their data-sharing preferences within a smart contract, and the contract ensures that data is shared only in alignment with these predefined terms.

Dynamic Consent Management: One of the defining features of smart contracts is their programmability. In the context of user consent, this translates into dynamic consent management. Users can modify their consent preferences in real time through the smart contract, enabling them to maintain control over their data-sharing decisions as circumstances change.

Transparent Execution: Smart contracts operate in a transparent and tamper-proof environment provided by the blockchain. This transparency ensures that the execution of consent agreements is verifiable by all parties, enhancing trust and accountability.

Self-Enforcing Agreements: Once the terms of consent are encoded into a smart contract, they become self-enforcing. The contract ensures that data is only shared or accessed according to the agreed-upon terms, minimizing the risk of unauthorized data use.

Immutable Record of Consent: Every transaction involving the execution of a smart contract is recorded on the blockchain. This creates an immutable record of consent events, enhancing transparency and providing an auditable trail of data-sharing decisions.

The utilization of smart contracts for consent management not only streamlines the process but also empowers users with greater agency over their data. This synergy between blockchain and IoT holds the potential to reshape how consent is managed and enforced in the digital age, enhancing privacy and promoting user-centric data governance.

10.6.3 Enhancing User Privacy with Blockchain–IoT Fusion

As the fusion of blockchain and the Internet of Things (IoT) unfolds, it brings about a paradigm shift in the way user privacy is safeguarded (Figure 10.6). This section delves into how this integration can significantly enhance user privacy within the digital landscape:

Decentralized Identity Management: Blockchain's decentralized nature enables users to maintain control over their identities. Instead of relying on centralized identity providers, users can manage their own identities.

FIGURE 10.6 User privacy with blockchain–IoT fusion.

Data Minimization: The integration of blockchain and IoT empowers users to share only the necessary data required for a specific transaction or interaction.

Selective Data-Sharing: Blockchain facilitates selective data-sharing, allowing users to share specific data points rather than entire datasets. This granular control ensures that only relevant information is shared, safeguarding sensitive details.

Immutable Audit Trail: Every interaction recorded on the blockchain creates an immutable audit trail. This transparency and traceability enhance user trust by ensuring that data handling adheres to predefined terms and is free from tampering.

Enhanced Consent Mechanisms: The integration of blockchain and IoT enables more robust and tamper-proof consent mechanisms through smart contracts. Users can define consent terms that are automatically enforced by the blockchain, ensuring that data is used only as authorized.

Combating Unauthorized Access: Blockchain's cryptographic techniques and consensus mechanisms make unauthorized data access significantly more challenging. This heightens the security of IoT-generated data, reducing the risk of breaches.

In essence, the fusion of blockchain and IoT transcends mere technological integration; it embodies a transformative shift in how user privacy is upheld in the digital landscape. By enabling decentralized identity management, fine-tuned data-sharing, and automated consent enforcement, this fusion sets the stage for a privacy-centric era of data interaction and collaboration.

10.7 SCALABILITY AND PERFORMANCE CONSIDERATIONS

10.7.1 Scalability Challenges in Both Blockchain and IoT

The integration of blockchain and the Internet of Things (IoT) holds great promise, yet it also presents significant scalability challenges for both technologies:

Scalability Challenges in IoT

Data Volume: The proliferation of IoT devices results in an enormous volume of data generated. This flood of information strains network bandwidth and processing capabilities, leading to potential bottlenecks.

Data Transmission: Transferring large amounts of data from IoT devices to the blockchain can cause delays and network congestion, affecting real-time responsiveness.

Resource Constraints: Many IoT devices operate with limited computational resources, making it challenging to perform complex tasks and manage cryptographic operations required by the blockchain.

Scalability Challenges in Blockchain

Transaction Throughput: Traditional blockchains face limitations in the number of transactions they can process per second. This can hinder the efficient handling of high-frequency data flows from IoT devices.

Consensus Overhead: Consensus mechanisms, while essential for maintaining blockchain security, introduce overhead that can hinder scalability. As more participants join the network, consensus-related computations increase.

Storage Requirements: Storing every transaction on the blockchain can result in rapidly growing storage requirements, making the blockchain less feasible for IoT data storage.

Latency: The time taken to confirm transactions in a blockchain can conflict with the real-time requirements of IoT applications, where immediate response is crucial.

Addressing these scalability challenges is paramount to fully realize the potential of blockchain–IoT integration. Strategies enhance both blockchain's transaction throughput and IoT's data management efficiency for achieving a harmonious fusion of these technologies.

10.7.2 Solutions for Scalability in Blockchain–IoT Integration

The integration of blockchain and the Internet of Things (IoT) necessitates innovative solutions to ensure optimal scalability. This section explores strategies that address scalability challenges within this dynamic fusion:

Off-Chain Processing: Off-chain solutions divert certain transactions away from the main blockchain, reducing congestion. This approach, often utilizing side chains or state channels, enhances transaction throughput and responsiveness.

Sharding: Sharding is the process of breaking up the blockchain into smaller, more manageable chunks called shards. Scalability is greatly increased and parallel processing is made possible by each shard processing a subset of transactions.

IoT Edge Computing: Adopting edge computing lowers the amount of data sent to the blockchain by processing data closer to Internet of Things devices. By doing this, system efficiency is increased overall and network congestion is reduced.

Optimized Data Aggregation: Reducing the amount of data by aggregating it at the source before sending it to the blockchain addresses network congestion and preserves data accuracy.

Consensus Algorithm Selection: Choosing consensus algorithms that prioritize scalability, such as Delegated Proof of Stake (DPoS) or Proof of Authority (PoA), can enhance transaction throughput while preserving security.

Hybrid Approaches: A combination of on-chain and off-chain processing can balance scalability and security. For example, on-chain settlement can be preceded by off-chain transactions, reducing blockchain load.

In navigating the landscape of scalability within blockchain–IoT integration, adopting these solutions ensures that the potential of this fusion is harnessed without compromising performance, responsiveness, or security.

10.7.3 Balancing Performance and Security

The synergy between blockchain and the Internet of Things (IoT) offers immense potential, but it requires a delicate equilibrium between performance and security considerations. This section examines the critical task of harmonizing these aspects within the context of blockchain–IoT integration:

Transaction Speed versus Security: Striking the right balance between transaction speed and security is pivotal. While faster transaction processing is desirable, it should not compromise the robustness of security mechanisms.

Resource Constraints: IoT devices often operate with limited computational resources. Balancing performance involves optimizing data processing while ensuring that resource limitations are not exceeded.

Consensus Mechanisms: Selecting appropriate consensus mechanisms are crucial. Some algorithms prioritize speed, while others emphasize security. Choosing the right mechanism ensures that both facets are adequately addressed.

Scalability Solutions: As discussed earlier, scalability solutions such as sharding and off-chain processing enhance performance. However, they must be implemented carefully to avoid introducing vulnerabilities.

Encryption and Authentication: Maintaining robust encryption and authentication mechanisms is essential. While they add a layer of security, they can also introduce processing overhead. Balancing their implementation with performance is crucial.

Real-Time Responsiveness: IoT applications often demand real-time responsiveness. Achieving this without compromising security requires careful orchestration of data flow and processing.

User Experience: Balancing performance and security impacts the overall user experience. Slow transactions or data delays can deter users, making optimization critical.

In the dynamic landscape of blockchain–IoT fusion, achieving the right equilibrium between performance and security is pivotal. A nuanced approach that considers the unique requirements of each application and leverages suitable technologies ensures that the integration reaps the benefits of both performance and security without undue compromise.

10.8 CASE STUDIES: REAL-WORLD BLOCKCHAIN–IoT IMPLEMENTATIONS

10.8.1 Industrial IoT and Supply Chain Management

In this section, we delve into real-world instances where the integration of blockchain and the Internet of Things (IoT) has revolutionized industrial processes and supply chain management:

Enhanced Traceability: The fusion of blockchain and IoT has empowered supply chain stakeholders with unparalleled traceability. For example, IBM

Food Trust employs this fusion to trace the journey of food products from farm to table, mitigating risks and ensuring authenticity.

Provenance Verification: Luxury goods brands like LVMH utilize blockchain–IoT integration to authenticate the origin of products. IoT sensors coupled with blockchain ensure that each step of the production process is verified and transparent.

Efficient Inventory Management: IoT sensors in warehouses and storage facilities monitor inventory levels. Blockchain integration automates inventory tracking, reducing human intervention and enhancing accuracy.

Counterfeit Prevention: Brands combat counterfeiting by integrating IoT sensors with blockchain. Every product is equipped with a unique identifier, and blockchain ensures the immutable recording of its journey, preventing counterfeit products from entering the market.

This case study illustrates how the amalgamation of blockchain and IoT in industrial IoT and supply chain management has led to enhanced transparency, security, and efficiency across the life cycle of products.

10.8.2 Healthcare and Patient Data Management

This section examines tangible applications where the convergence of blockchain and the Internet of Things (IoT) has reshaped healthcare and patient data management (Figure 10.7):

Secure Medical Records: Blockchain–IoT integration ensures the security and integrity of patient medical records. Estonia's eHealth system employs this fusion to allow patients to control access to their health data securely.

Clinical Trials and Research: IoT devices collect real-time data from patients participating in clinical trials. Blockchain guarantees the transparency and authenticity of this data, fostering trust between participants and researchers.

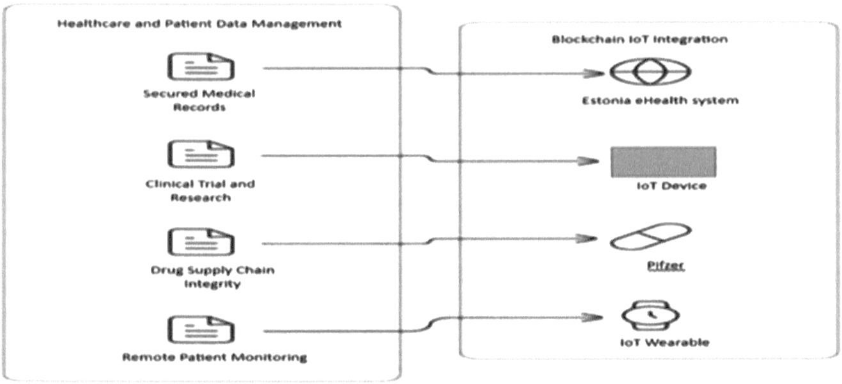

FIGURE 10.7 Healthcare and patient data management.

Drug Supply Chain Integrity: Blockchain–IoT integration combats the proliferation of counterfeit drugs. Pharmacy giants like Pfizer use this fusion to track the journey of drugs through the supply chain, ensuring authenticity.

Remote Patient Monitoring: IoT wearables enable remote monitoring of patients' health parameters. Integration with blockchain guarantees the immutability of this data, ensuring trustworthiness for medical professionals.

This case study exemplifies the transformative potential of merging blockchain and IoT in the healthcare sector. By enhancing data security, interoperability, and patient-centric control, this fusion paves the way for more efficient and patient-focused healthcare systems.

10.8.3 Energy Management and Smart Grids

In this section, we explore real-world examples where the fusion of blockchain and the Internet of Things (IoT) has reshaped energy management and smart grid systems:

Decentralized Energy Trading: The integration of blockchain and IoT facilitates peer-to-peer energy trading. Platforms like Power Ledger enable consumers to buy and sell excess energy directly, enhancing energy efficiency and reducing costs.

Grid Optimization: IoT sensors embedded in smart grids continuously monitor energy consumption and grid health. Blockchain integration ensures the integrity of these sensor readings, contributing to accurate grid management and optimization.

Renewable Energy Tracking: Blockchain–IoT fusion enhances transparency in renewable energy generation. SolarCoin leverages this integration to incentivize solar energy production by recording each megawatt-hour of generated solar power on the blockchain.

Microgrids and Resilience: In scenarios where the main grid fails, microgrids can operate autonomously. IoT devices monitor microgrid parameters, and blockchain ensures that these devices data is secure and tamper-proof.

These case studies underscore the transformative impact of merging blockchain and IoT in energy management and smart grids. By enabling efficient peer-to-peer energy trading, optimizing grid performance, and fostering transparency in renewable energy, this fusion accelerates the transition toward a sustainable and decentralized energy future.

10.9 CHALLENGES AND LIMITATIONS

10.9.1 Interoperability Issues in Heterogeneous IoT Devices

Integrating blockchain and the Internet of Things (IoT) introduces a complex array of devices with varying protocols, standards, and capabilities. This section delves into the challenges posed by the heterogeneity of IoT devices:

Diverse Communication Protocols: IoT devices often utilize different communication protocols, making it difficult to establish seamless connectivity and data exchange within a blockchain network.

Data Formats and Standards: IoT devices generate data in various formats, complicating data aggregation and interpretation. Standardizing data formats is crucial for coherent data processing and utilization.

Scalability across Devices: Ensuring scalability across diverse IoT devices is challenging. Integrating a vast number of devices while maintaining efficient transaction processing requires careful orchestration.

Legacy Systems Integration: Incorporating legacy IoT devices into a blockchain network is intricate, as these devices may lack the necessary compatibility with blockchain protocols.

Solution-Driven Approaches: Overcoming interoperability challenges demands collaborative efforts to develop standardized communication protocols, data formats, and middleware that bridge the gap between various IoT devices. A cohesive approach ensures that blockchain and IoT can synergize effectively across different device types, maximizing their combined potential.

10.9.2 Energy Efficiency in Blockchain–IoT Networks

The convergence of blockchain and the Internet of Things (IoT) presents energy efficiency challenges that require careful consideration:

Increased Energy Consumption: Integrating them can amplify energy consumption, impacting the operational costs and environmental sustainability of the system.

Proof-of-Work (PoW) Limitations: Traditional blockchain networks using PoW consensus can be particularly energy-intensive. The energy consumed during mining processes raises concerns about the ecological footprint of blockchain–IoT fusion.

IoT Device Constraints: Many IoT devices operate with limited energy resources. Integrating energy-efficient protocols and mechanisms into blockchain–IoT networks is crucial to extend the operational lifespan of these devices.

Hybrid Approaches: Exploring alternative consensus mechanisms like Proof of Stake (PoS) or Delegated Proof of Stake (DPoS) can significantly reduce energy consumption in blockchain networks, making them more amenable to IoT integration.

Optimization Strategies: Employing techniques like off-chain processing, where certain transactions occur off the main blockchain, can reduce the energy burden on the blockchain network.

Addressing the energy efficiency challenge demands a holistic approach. By adopting energy-efficient consensus mechanisms, optimizing transaction processing, and prioritizing low-power IoT device integration, the blockchain–IoT fusion can be energy-conscious without sacrificing functionality.

10.9.3 Regulatory and Legal Implications

The integration of blockchain and the Internet of Things (IoT) gives rise to a complex web of regulatory and legal considerations that require meticulous navigation:

Data Privacy and Ownership: The collection and sharing of data through IoT devices must comply with data protection regulations, such as the General Data Protection Regulation (GDPR). Determining data ownership and consent mechanisms becomes crucial.

Cross-Border Data Transfer: IoT devices often transmit data across borders, triggering compliance with diverse international data protection laws. Ensuring seamless cross-border data transfer while adhering to regulations is a challenge.

Smart Contracts and Legal Validity: The automation facilitated by smart contracts raises questions about their legal validity and enforceability. Striking a balance between code-based execution and legal compliance is essential.

Liability and Accountability: In the event of malfunction or misbehavior of IoT devices, attributing liability can be complex. The integration necessitates clear delineation of responsibilities among stakeholders.

Regulatory Alignment: Blockchain and IoT often span multiple industries, each with its regulatory framework. Harmonizing these regulations to facilitate blockchain–IoT integration can be intricate.

Legal Frameworks: Crafting legal frameworks that encompass the unique challenges of blockchain–IoT integration requires collaboration between legal experts, technologists, and policymakers.

In navigating these regulatory and legal complexities, a multidisciplinary approach that includes legal experts, technologists, and policymakers is indispensable. Ensuring compliance, data privacy, and legal validity are integral to the successful and ethical implementation of blockchain–IoT fusion.

10.10 FUTURE TRENDS AND RESEARCH DIRECTIONS

10.10.1 Edge Computing and Hybrid Architectures

The future evolution of the integration between blockchain and the Internet of Things (IoT) holds promising directions, particularly in the realm of edge computing and hybrid architectures:

Edge Computing Proliferation: The widespread adoption of edge computing is expected, enabling data processing closer to IoT devices. Exploring how blockchain can enhance security and data integrity in edge environments will be a pivotal research area.

Decentralized Edge Networks: Researching the implementation of decentralized blockchain networks at the edge can lead to more resilient and responsive IoT ecosystems, addressing latency concerns.

Hybrid Models for Scalability: The development of hybrid architectures that seamlessly combine on-chain and off-chain processing can enhance scalability and transaction efficiency while maintaining security and decentralization.

Efficiency and Sustainability: Investigating how edge computing can optimize energy consumption within blockchain–IoT networks is vital for building sustainable and eco-friendly solutions.

As the convergence of blockchain and IoT continues to mature, research and innovation in edge computing and hybrid architectures will play a central role in shaping the future landscape of secure and efficient data management.

10.10.2 Integration with AI and Machine Learning

The trajectory of the integration between blockchain and the Internet of Things (IoT) is poised to intersect with the realms of AI and machine learning, offering novel avenues for research and development:

Advanced Analytics: Leveraging algorithms within networks can enable advanced data analysis, yielding insights that were previously unattainable.

Predictive Modeling: Integrating predictive modeling into blockchain–IoT systems empowers real-time decision-making by anticipating trends and potential issues based on historical data.

Smart Contract Optimization: AI can be utilized to optimize smart contract execution, ensuring efficient resource allocation and adherence to predefined conditions.

Data Anomaly Detection: Machine learning algorithms can identify anomalies in IoT data streams, enhancing security and early detection of irregularities.

Exploring the synergies between blockchain, IoT, AI, and machine learning will not only amplify the potential of each technology but also pave the way for innovative solutions that revolutionize industries and domains.

10.10.3 Standardization Efforts and Collaborative Research

The integration of blockchain and the Internet of Things (IoT) is poised for impactful growth, driven by standardization efforts and collaborative research:

Interoperable Protocols: Establishing standardized communication protocols and data formats for IoT devices across various industries will foster seamless integration with blockchain networks.

Global Regulatory Alignment: Collaborative efforts to align regulatory frameworks internationally will facilitate the deployment of blockchain–IoT solutions across borders while ensuring compliance.

Cross-Disciplinary Collaboration: Collaborative research involving academia, industry, legal experts, and policymakers will enable holistic solutions that address technical, ethical, and legal aspects.

Ecosystem Building: Building open and collaborative ecosystems that encourage the development and sharing of best practices will accelerate innovation in blockchain–IoT fusion.

This fusion gains momentum, dedicating research efforts to standardization and collaboration is paramount. These efforts will lay the foundation for a harmonized and robust integration that spans industries, promotes innovation, and addresses challenges effectively.

10.11 CONCLUSION

10.11.1 Recap of Blockchain–IoT Integration Benefits

In conclusion, the integration of blockchain and the Internet of Things (IoT) offers a multitude of benefits that have the potential to reshape various industries and domains:

In healthcare, it enables secure and interoperable patient data management, empowering individuals to have control over their health information.

For energy management, the integration optimizes the distribution of resources, contributing to more sustainable and efficient grids.

The convergence of these technologies holds the promise of not only enhancing operational efficiency but also revolutionizing business models and user experiences across sectors.

In essence, empowering organizations to achieve new heights of data-driven excellence and transformative impact.

10.11.2 Call to Action for Industry and Research Community

In closing, the convergence of blockchain and the Internet of Things (IoT) holds tremendous potential for innovation and transformative change. As we wrap up our exploration of this dynamic fusion, a call to action resonates strongly for both the industry and the research community:

Industry Engagement: Embrace the collaborative spirit by actively participating in the development of interoperable protocols and standards for IoT devices, ensuring seamless integration with blockchain networks. Invest in research and development to explore novel use cases and practical implementations that harness the combined power of blockchain and IoT to solve real-world challenges.

Research Community Involvement: Continue to drive innovation by delving into the technical intricacies of blockchain–IoT integration, exploring new consensus mechanisms, data management strategies, and security protocols.

Collaborate across disciplines, engaging legal experts, policymakers, and ethicists to address the multifaceted challenges presented by this fusion. Together, through industry commitment and research innovation, we can shape a future where security, transparency, and efficiency converge to redefine industries and the way we interact with technology.

REFERENCES

1. Xiong, Z., Zhang, Y., Luong, N. C., Niyato, D., Wang, P., & Guizani, N. (2020). The best of both worlds: A general architecture for data management in blockchain-enabled internet-of-things. *IEEE Network*, 34(1), 166–173. https://doi.org/10.1109/MNET.001.1900095
2. Shafagh, H., Burkhalter, L., Hithnawi, A., & Duquennoy, S. (2017, November). Towards blockchain-based auditable storage and sharing of IoT data. In *Proceedings of the 2017 on Cloud Computing Security Workshop* (pp. 45–50). https://doi.org/10.1145/3140649.3140656
3. Novo, O. (2018). Blockchain meets IoT: An architecture for scalable access management in IoT. *IEEE Internet of Things Journal*, 5(2), 1184–1195. https://doi.org/10.1109/JIOT.2018.2812239
4. Pinno, O. J. A., Gregio, A. R. A., & De Bona, L. C. (2017, December). Controlchain: Blockchain as a central enabler for access control authorizations in the IoT. In *GLOBECOM 2017–2017 IEEE Global Communications Conference* (pp. 1–6). IEEE. https://doi.org/10.1109/GLOCOM.2017.8254521
5. Luong, N. C., Xiong, Z., Wang, P., & Niyato, D. (2018, May). Optimal auction for edge computing resource management in mobile blockchain networks: A deep learning approach. In *2018 IEEE International Conference on Communications (ICC)* (pp. 1–6). IEEE. https://doi.org/10.1109/ICC.2018.8422743
6. Datiri, D. D., & Li, M. (2023, June). A cluster enabled blockchain-based data management for IoT systems. In *2023 24th International Carpathian Control Conference (ICCC)* (pp. 88–92). IEEE. https://doi.org/10.1109/ICCC57093.2023.10178949
7. Okegbile, S. D., Cai, J., & Alfa, A. S. (2022). Performance analysis of blockchain-enabled data-sharing scheme in cloud-edge computing-based IoT networks. *IEEE Internet of Things Journal*, 9(21), 21520–21536. https://doi.org/10.1109/JIOT.2022.3181556
8. Yang, C., Jiang, P., & Zhu, L. (2022). Accelerating decentralized and partial-privacy data access for VANET via online/offline functional encryption. *IEEE Transactions on Vehicular Technology*, 71(8), 8944–8954. https://doi.org/10.1109/TVT.2022.3174888
9. Naghib, A., Jafari Navimipour, N., Hosseinzadeh, M., & Sharifi, A. (2023). A comprehensive and systematic literature review on the big data management techniques in the internet of things. *Wireless Networks*, 29(3), 1085–1144. https://doi.org/10.1007/s11276-022-03177-5
10. Tian, Y. (2022). AI-assisted dynamic modeling for data management in a distributed system. *Journal of Interconnection Networks*, 22(Supp05), 2147002. https://doi.org/10.1142/S0219265921470022
11. Shen, X. S., Liu, D., Huang, C., Xue, L., Yin, H., Zhuang, W.,. . . Ying, B. (2022). Blockchain for transparent data management toward 6G. *Engineering*, 8, 74–85. https://doi.org/10.1016/j.eng.2021.10.002
12. Peng, S., Zhu, L., Cai, Z., Liu, W., He, C., & Tang, W. (2021). Dynamic optimization of government data transmission based on blockchain technology. *Mobile Information Systems*, 2021, 1–8. https://doi.org/10.1155/2021/8948323
13. Bali, V., Bali, S., Gaur, D., Rani, S., & Kumar, R. (2023). Commercial-off-the shelf vendor selection: A multi-criteria decision-making approach using intuitionistic

fuzzy sets and TOPSIS. *Operational Research in Engineering Sciences: Theory and Applications*, 6(2).
14. Kumar, P., Banerjee, K., Singhal, N., Kumar, A., Rani, S., Kumar, R., & Lavinia, C. A. (2022). Verifiable, secure mobile agent migration in healthcare systems using a polynomial-based threshold secret sharing scheme with a blowfish algorithm. *Sensors*, 22(22), 8620.
15. Rani, S., Kataria, A., & Chauhan, M. (2022). Cyber security techniques, architectures, and design. In *Holistic Approach to Quantum Cryptography in Cyber Security* (pp. 41–66). USA: CRC Press.
16. Chauhan, M., & Rani, S. (2021). Covid-19: A revolution in the field of education in India. *Learning How to Learn Using Multimedia*, 23–42.
17. Rani, S., Pareek, P. K., Kaur, J., Chauhan, M., & Bhambri, P. (2023, February). Quantum machine learning in healthcare: Developments and challenges. In *2023 IEEE International Conference on Integrated Circuits and Communication Systems (ICICACS)* (pp. 1–7). Raichur: IEEE.
18. Puri, V., Kataria, A., Solanki, V. K., & Rani, S. (2022, December). AI-based botnet attack classification and detection in IoT devices. In *2022 IEEE International Conference on Machine Learning and Applied Network Technologies (ICMLANT)* (pp. 1–5). Slvador: IEEE.
19. Sudevan, S., Barwani, B., Al Maani, E., Rani, S., & Sivaraman, A. K. (2021). Impact of blended learning during Covid-19 in Sultanate of Oman. *Annals of the Romanian Society for Cell Biology*, 14978–14987.
20. Kataria, A., Agrawal, D., Rani, S., Karar, V., & Chauhan, M. (2022). Prediction of blood screening parameters for preliminary analysis using neural networks. In *Predictive Modeling in Biomedical Data Mining and Analysis* (pp. 157–169). New York, NY: Academic Press.
21. Rani, S., Bhambri, P., & Kataria, A. (2023). Integration of IoT, big data, and cloud computing technologies. In *Big Data, Cloud Computing and IoT: Tools and Applications*. USA: CRC Press.
22. Tanwar, R., Chhabra, Y., Rattan, P., & Rani, S. (2022, September). Blockchain in IoT networks for precision agriculture. In *International Conference on Innovative Computing and Communications: Proceedings of ICICC 2022* (Vol. 2, pp. 137–147). Singapore: Springer Nature Singapore.
23. Bhambri, P., & Rani, S. (2024). Challenges, opportunities, and the future of industrial engineering with IoT and AI. In *Integration of AI-Based Manufacturing and Industrial Engineering Systems with the Internet of Things* (pp. 1–18). USA: CRC Press.
24. Bhambri, P., Rani, S., Balas, V. E., & Elngar, A. A. (Eds.). (2023). *Integration of AI-Based Manufacturing and Industrial Engineering Systems with the Internet of Things*. USA: CRC Press.
25. Rani, S., Kaur, J., & Bhambri, P. (2023). Technology and gender violence: Victimization model, consequences and measures. In *Communication Technology and Gender Violence* (pp. 1–19). Cham: Springer International Publishing.
26. Singh, H., & Rani, S. (2023). Flex sensor integrated smart strap to verify correct wearing of the face mask. *IEEE Sensors Journal*, 24(2), 2020–2027.
27. Rani, S., Kumar, S., Kataria, A., & Min, H. (2023). SmartHealth: An intelligent framework to secure IoMT service applications using machine learning. *ICT Express*, 10(2), 420–425.
28. Rani, S., Mishra, A. K., Kataria, A., Mallik, S., & Qin, H. (2023). Machine learning-based optimal crop selection system in smart agriculture. *Scientific Reports*, 13(1), 15997.

29. Kataria, A., Puri, V., Pareek, P. K., & Rani, S. (2023, July). Human activity classification using G-XGB. In *2023 International Conference on Data Science and Network Security (ICDSNS)* (pp. 1–5). Tiptur: IEEE.
30. Rani, S., Kataria, A., Kumar, S., & Tiwari, P. (2023). Federated learning for secure IoMT-applications in smart healthcare systems: A comprehensive review. *Knowledge-Based Systems*, 110658.

11 Empowering IoT Security

Enhancing Applications through Edge Computing and Blockchain

Mahfooz Alam, Aejaz Nazir Lone, and Suhel Mustajab

11.1 INTRODUCTION

The Internet of Things (IoT) paradigm enables the interconnection of several different kinds of machines, sensors, and devices, allowing them to interact and transmit data over the internet. This integration facilitates the collection and analysis of data through applications, leading to the identification of events and the activation of actions via actuators (Rani et al., 2022). The result is the development of cyber-physical systems with minimal human intervention in sectors like healthcare, transportation, utilities, agriculture, and surveillance. This advancement paves the way for the creation of smarter environments such as smart cities, health systems, transportation networks, power grids, and security systems, all capable of making real-time decisions and improving overall efficiency (Alam & Siddiqui, 2023; Lone et al., 2023).

Until recently, on-demand resource sharing had to rely on the highly flexible and scalable nature of traditional cloud computing (Kumar et al., 2022). However, with the recent exponential expansion of IoT, increased real-time demands for data storage, processing, and exchange have emerged that are much above the maximum capacity of conventional cloud computing. A brand-new computing paradigm called edge computing has been put out as a significant addition to cloud computing in order to bring additional processing and storage resources from distant clouds to the edge of the IoT. By doing so, the computational, storage, and bandwidth demands of conventional cloud computing may be substantially alleviated by using the resources of edge devices. Edge computing ensures privacy protection and data security by offloading computing functions to edge devices that are closer to data sources (Chen et al., 2018). However, many of the current data-security solutions cannot be used to their full potential in the edge computing architecture because of the restricted resources of edge devices, the heterogeneity of networks, and the highly dynamic environment (Xiao et al., 2019).

Recently, blockchain has received much interest as a potential alternative approach to enhance the safety and effectiveness of edge computing (Wu et al.,

DOI: 10.1201/9781003460367-11

2020). Blockchain may be thought of as a decentralized ledger that employs P2P, cryptographic, distributed storage, and other technologies to accomplish decentralization, transparency, traceability, security, and immutability. By storing sensitive data on blockchain, it can, in essence, improve the security and integrity of edge nodes in edge computing (Luo et al., 2021). With the use of carefully designed smart contracts, blockchain technology can also allow for the deployment of other security measures in edge computing, such as access control, authentication, and the protection of user privacy. In turn, blockchain may be supported by edge computing by offering sufficient processing power for mining operations. Because of the complementary nature of the two technologies, combining edge computing with blockchain offers a viable paradigm for building frameworks to address issues in a wide range of domains (Rani, Bhambri, et al., 2023).

11.1.1 Overview of IoT Security Challenges

The Internet of Things (IoT), which connects various kinds of devices and systems to improve daily life, has completely changed the way we interact with the outside world. However, there are a number of security issues that come with this interconnection. Due to the limitations of many IoT devices in terms of their computing, power, and communication capacities, the majority of legacy security protocols used on the conventional internet cannot be directly translated to the IoT scenario (Caparra et al., 2017).

The diversity of IoT devices and their varying levels of security features further compounds the challenge. IoT networks are vulnerable to hacking, with potential risks ranging from data breaches and unauthorized access to device manipulation and disruption of critical services. The lack of standardized security protocols and the rapid pace of IoT development often lead to vulnerabilities, leaving devices exposed to attacks. Furthermore, issues such as weak authentication, inadequate encryption, and insufficient firmware updates present additional hurdles in securing IoT ecosystems. Addressing these challenges necessitates a comprehensive approach that encompasses device authentication, data encryption, secure communication protocols, and regular security updates to ensure a robust and resilient IoT infrastructure.

11.1.2 Contributions of the Study

This study contributes to the field of IoT security by providing a comprehensive exploration of IoT security challenges. It explores the potential of edge computing to improve IoT security by bringing data processing and analytics closer to the network edge, enabling local threat detection, real-time responses, and data filtering. It also explores how blockchain technology could enhance data integrity, device authentication, and secure communication in IoT ecosystems by offering decentralized security and trust. The combination of edge computing and blockchain is further explored in the study, with a focus on how their combined advantages might improve IoT security. It addresses challenges and considerations in implementing these technologies and identifies future research opportunities.

11.1.3 Chapter Organization

The main aim of this chapter is to explore the convergence of edge computing and blockchain technologies for enhancing IoT security. A general overview of IoT security challenges is provided in Section 11.1. Section 11.1.2 provides the contributions of the study. A brief discussion about major security attacks is given in Section 11.2. Section 11.2.1 discusses threats and vulnerabilities in IoT. Section 11.2.2 discusses authentication and access control issues. Section 11.2.3 discusses data privacy and confidentiality concerns, and Section 11.2.4 discusses scalability and performance challenges. Edge computing for IoT security is discussed in Section 11.3. Section 11.4 discussed blockchain and its advantages, which prohibit unauthorized access, data tampering, and identity spoofing. Section 11.5 discusses the motivation for combining edge computing and blockchain to enhance each other's security. In Section 11.6, we discuss how the combined edge computing and blockchain technology approach can make the IoT system more secure. Possible challenges and limitations of the combined approach are provided in Section 11.7. Section 11.8 discusses the opportunities and possible future trends in the IoT world. The chapter concludes in Section 11.9.

11.2 IoT SECURITY CHALLENGES

In this section, some of the essential IoT security challenges have been described.

11.2.1 Threats and Vulnerabilities in IoT Systems

The Internet of Things (IoT) faces various threats and vulnerabilities that can compromise its security (Lone et al., 2023). Some common risks include unauthorized access, data privacy breaches, denial-of-service attacks, physical tampering, man-in-the-middle attacks, reply attacks, malware, and inadequate firmware security (Bali et al., 2023). Additionally, challenges like lack of security updates and standardization issues contribute to the vulnerability of IoT systems. To address these risks, a multilayered approach is necessary. In Puthal et al. (2016), the authors suggested a framework that can identify reply attacks and take precautions against them. Haddad Pajouh et al. (2018) suggested a methodology that makes use of a deep learning model for detecting IoT malware. Measures such as strong authentication, data encryption, regular patching, secure coding, network segmentation, intrusion detection systems, user awareness, and education can help mitigate these risks and enhance IoT security. It is also important to comply with standards, ensure physical security, monitor for incidents, and manage vendor risks.

11.2.2 Authentication and Access Control Issues

Authentication and access control are critical aspects of IoT security. As IoT systems connect numerous devices and handle sensitive data, it is crucial to ensure that only authorized entities can access and interact with IoT devices and their data. Before communicating with the system or other devices on the IoT platform, each and every device must first be identified and authenticated (Fati et al., 2018). Weak authentication mechanisms, lack of access control policies, and insecure communication channels are some common issues. To address these challenges, strong authentication methods, robust identity management systems, well-defined access control policies, and secure communication channels should be implemented. By following these practices, IoT systems can maintain integrity, confidentiality, and protection against evolving threats (Chauhan & Rani, 2021).

11.2.3 Data Privacy and Confidentiality Concerns

In the IoT, data privacy and confidentiality are major concerns due to the large amounts of sensitive data generated, transmitted, and processed by IoT devices. Providing data access to authorized individuals or entities is a crucial task. Data management is essential for maintaining data security and ensuring that only authorized devices and objects have access to it (Qadri et al., 2021). Key concerns include sensitive data exposure, inadequate data encryption, lack of data minimization, insecure data storage, third-party data-sharing, privacy in data analytics, lack of user consent and control, and privacy in cross-domain integration. To address these concerns, it is important to implement strong data encryption, data minimization practices, secure storage and transmission mechanisms, privacy-aware data analytics, user-centric privacy controls, and compliance with privacy regulations. Privacy should be considered throughout the entire life cycle of IoT systems to ensure data is handled in a privacy-preserving manner (Bhambri et al., 2023).

11.2.4 Scalability and Performance Challenges

Scalability and performance issues are critical challenges in IoT due to the massive scale and diverse requirements of IoT systems. The key challenges include handling and processing large volumes of data in real time, managing network congestion and limited bandwidth, overcoming resource constraints of IoT devices, scaling device management systems, ensuring interoperability and standards, achieving real-time processing and response, and balancing security and privacy with performance. Addressing these challenges requires architectural considerations, efficient data management, optimized networks, standardized protocols, and hardware advancements. Continuous monitoring and optimization are essential for successful large-scale IoT deployments. In a survey on the IoT, Kouicem et al. (2018) found that heterogeneity, scalability, mobility, and resource constraints were the main security issues. The authors also highlighted that new, cutting-edge methods, such as those that employ SDN and blockchain technology, may be utilized to address security challenges in IoT more effectively.

11.3 EDGE COMPUTING FOR IoT SECURITY

The term "edge computing" refers to an innovative form of distributed computing that places the storage and processing of data closer to the network's periphery (Satyanarayanan et al., 2015). It processes data and performs computation closer to where it is generated, which is especially beneficial for real-time or near-real-time analysis. An edge computing system includes the internet gateway, edge servers, and perception layer. The internet gateway is responsible for connecting the edge system with the network backbone and cloud servers. Edge servers process, store, and aggregate data. The perception layer includes edge devices such as IoT sensors and actuators, smartphones, wearable devices, and personal computers. Edge computing and edge storage capabilities add new dimensions to network performance optimization problems and enlarge their feasible region. The operator can dynamically decide, depending on the situation, how the edge servers come into play and bring advantages to the whole network. For example, edge servers schedule computing tasks (Chen et al., 2021), provide better coverage, run load-balancing algorithms, resolve computation bottlenecks (Mijuskovic et al., 2021), and provide heterogeneous services well addressed in the MEC (multi-access edge computing) standard. This methodology facilitates computational offloading with the least amount of energy use. Figure 11.1 depicts the concept of edge computing in IoT systems. It emphasizes a network of interconnected edge devices, processing data locally, and enhancing

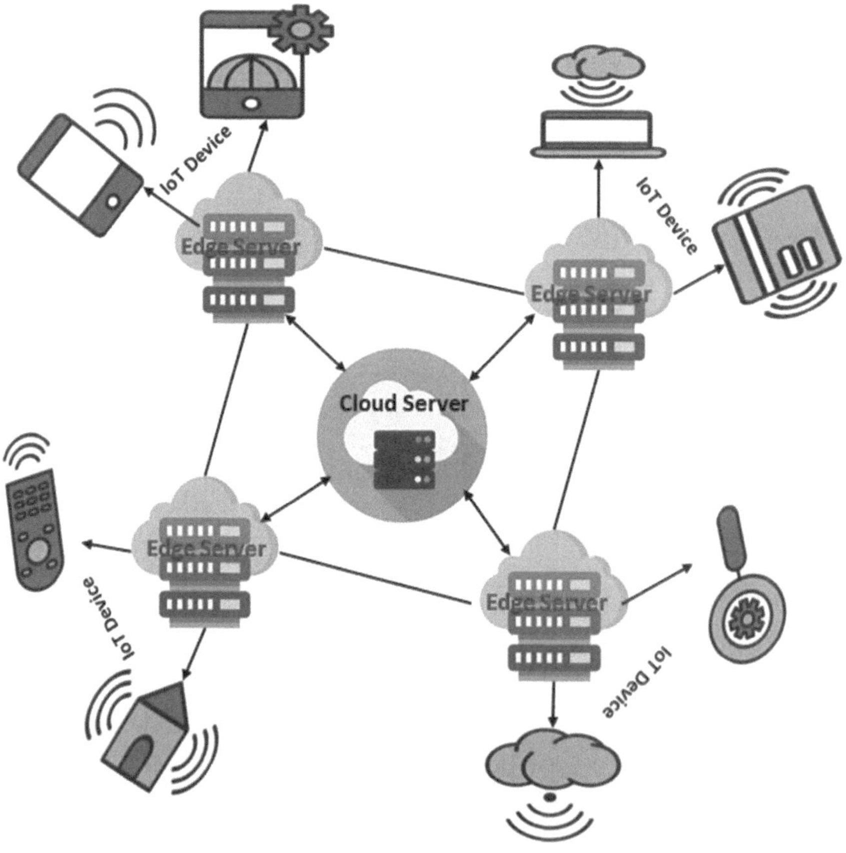

FIGURE 11.1 Edge computing paradigm.

security. The proximity of edge computing enables real-time analysis, reduced data transmission, and improved privacy, offering a resilient and secure infrastructure for IoT applications.

With the rise of edge computing, specific security measures for edge environments, such as secure bootstrapping, data encryption, and secure communication protocols, have become important. Zero Trust Architecture is an approach that assumes no implicit trust and focuses on continuous authentication and strict access controls. Edge-based anomaly detection and intrusion detection systems (IDS) are security solutions deployed at the network edge to detect and prevent anomalous or malicious activities. These systems analyze network traffic patterns, system logs, and other relevant data to identify deviations from normal behavior or signs of intrusion attempts. By operating closer to the source of data or communication, they offer real-time detection, reduced network latency, bandwidth optimization, distributed defense, and privacy protection (Kataria et al., 2022).

In the context of IoT systems, edge computing plays a crucial role in enhancing network security by providing faster response times, minimizing network latency, optimizing bandwidth usage, and offering a decentralized and redundant approach to security. It provides advantages such as data privacy, reduced attack surface, real-time threat detection, bandwidth optimization, and distributed security. However, challenges like local vulnerabilities, device authentication, and secure software updates need to be addressed for robust IoT security.

11.4 BLOCKCHAIN FOR IoT SECURITY

Blockchain technology is a decentralized and transparent digital ledger system that allows multiple parties to securely record and verify transactions. It operates on a network of computers called nodes, where each transaction is recorded in a block linked together to form an immutable chain (Tanwar et al., 2022, September). Blockchain offers features such as decentralization, transparency, security through cryptography, and consensus mechanisms to validate transactions. Smart contracts, which are self-executing contracts stored on the blockchain, enable automation and trust. Without the need for a trusted third party or centralized authority, blockchain technology allows applications and users to freely share and access IoT data (Majeed et al., 2021). Blockchain's distributed ledger makes it possible to trace data in the network and validate transaction records (Deepa et al., 2022). The data is consequently resistant to modification attacks since the majority of nodes in the blockchain are required to agree before any alterations to the information within the network can be made (Xu et al., 2021). By utilizing cryptographic signatures and hashes, blockchain verifies and authenticates data, ensuring its integrity and origin. Blockchain facilitates secure and trustless transactions between IoT devices through smart contracts, eliminating intermediaries and thus reducing fraud (Miglani et al., 2020). It ensures data integrity by making data tamper-proof and transparent through timestamping and cryptographic linking. Every action is documented as a transaction on a blockchain, and each block includes a cryptographic hash of the one before it, as well as its metadata, a date, and the transactions (Luo et al., 2021), as shown in Figure 11.2. As a result, changing the data in one block will change the data in all the other blocks as

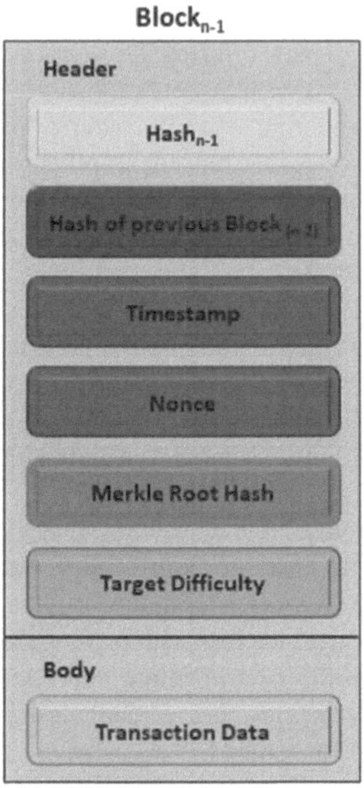

FIGURE 11.2 Blockchain for IoT security.

well which is impossible in the case of a large blockchain, thus making the block's data impervious to manipulation (Zhang et al., 2021).

Blockchain facilitates secure authentication and identity management for devices, allowing only authorized interactions (Sudevan et al., 2021). Blockchain's authentication, access control, and consensus protocols ensure that users are the sole owners of their data, and also simultaneously provide users with the ultimate privacy and security for their data (Guo & Yu, 2022). In summary, blockchain technology can enhance IoT security by providing a decentralized and distributed ledger, ensuring data integrity through cryptographic techniques, maintaining the immutability of recorded data, enabling authentication and identity management, supporting the execution of smart contracts, employing consensus mechanisms for validation, and offering privacy-enhancing features. These advantages address security challenges such as unauthorized access, data tampering, and identity spoofing. However, considerations should be made regarding scalability, energy consumption, and implementation complexities when implementing blockchain solutions for IoT security (Puri et al., 2022, December).

11.5 MOTIVATION FOR COMBINING EDGE COMPUTING AND BLOCKCHAIN TECHNOLOGIES

While edge computing significantly improves network performance by processing data closer to its source, it still faces security challenges such as authentication and potential malicious attacks. On the other hand, blockchain technology offers enhanced security, data integrity, and decentralized trust, but it grapples with limitations like scalability, energy consumption, storage requirements, and bandwidth strain (Rani, Kaur, et al., 2023). The motivation for combining these two technologies is due to the following two reasons.

11.5.1 Security Challenges in Edge Computing

Edge computing networks pose significant security challenges due to their distributed nature and interweaving of multiple technologies. Attacks such as jamming, sniffing, and data manipulation can disrupt communication and compromise data privacy. Ensuring the trustworthiness and validation of network configurations is crucial in this dynamic environment (Kataria et al., 2023, July). Managing heterogeneous networks becomes complex, making it difficult to isolate management traffic from regular data traffic, enabling adversaries to control the network more easily. Data integrity, privacy, and reliability are major concerns, as data is distributed across various storage locations. Additionally, secure computation and verifiability of results are essential (Rani, Kumar, et al., 2023).

11.5.2 Technical Challenges and Limitations of Blockchain

Blockchain technology faces challenges in scalability, storage requirements, interoperability, and energy consumption, which limit its widespread usage. The increasing storage space required as transactions grow and the limited throughput of public blockchains like Bitcoin and Ethereum hinder practical feasibility. Blockchain's decentralization is at risk if only a few entities can afford to run full nodes. Nevertheless, the expenses associated with the mining procedure, including the hardware cost and the power consumption needed for the CPU calculations, cannot be ignored (Maughelli, 2016). Combining edge computing and blockchain can mitigate some challenges by offloading computational tasks and improving efficiency in IoT systems (Singh & Rani, 2023).

11.6 COMBINING EDGE COMPUTING AND BLOCKCHAIN FOR ENHANCED IoT SECURITY

The emergence of new computing concepts such as big data, blockchain, edge computing, and IoT has introduced and facilitated us to perform resource-intensive computation, store data, and access information remotely (Dedeoglu et al., 2019). By combining the strengths of edge computing and blockchain technology, IoT systems benefit from improved security, trust, data integrity, and decentralized governance. Edge computing enables the processing and analysis of data closer to the source, reducing the need to send sensitive data to centralized cloud servers. By leveraging blockchain's decentralized and immutable nature, data can be securely stored and accessed without compromising privacy. Blockchain's cryptographic algorithms and consensus mechanisms provide a robust framework for ensuring data integrity and confidentiality (Rani, Pareek, et al., 2023, February). The combination of edge computing and blockchain technology offers scalability and flexibility for IoT

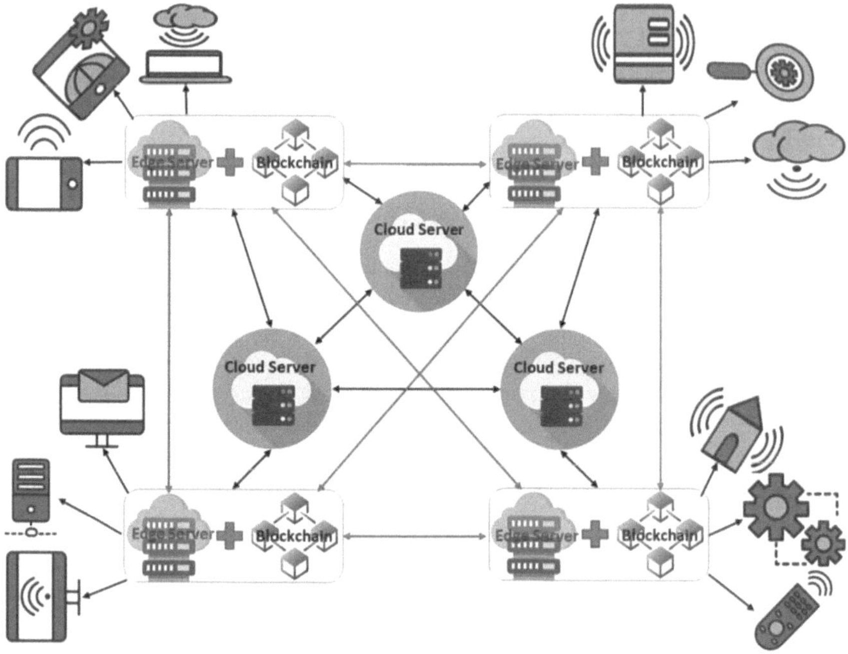

FIGURE 11.3 Integrated edge computing and blockchain for IoT security.

deployments. Edge computing allows for distributed processing and storage capabilities, enabling IoT devices to offload computation and storage tasks to nearby edge devices. Blockchain technology provides a scalable and decentralized framework for managing IoT transactions and maintaining a trusted ledger. This combination enables IoT systems to handle a large number of devices and transactions efficiently while maintaining security and integrity.

Figure 11.3 illustrates the integration of edge computing and blockchain in an IoT ecosystem. It depicts a network architecture where edge devices collect data from IoT devices and perform localized data processing. The processed data is then securely transmitted and stored on a blockchain network, ensuring decentralized security and immutable data integrity (Rani, Mishra, et al., 2023).

Also, centralized systems can be vulnerable to a single point of failure and malicious attacks. By integrating edge computing and blockchain, IoT networks can be decentralized, reducing the dependency on a single entity or server (Rani, Kataria, et al., 2023). This decentralization enhances the resilience and robustness of IoT systems, as there is no single point that can be targeted or compromised. The decentralized nature of blockchain allows for consensus mechanisms and distributed ledgers, ensuring data integrity and trust among multiple parties in an IoT ecosystem. Combining edge computing with blockchain can provide enhanced resilience to various types of attacks. Since data processing and storage occur at the edge, it reduces the attack surface and limits the impact of potential breaches. Even if an individual device is compromised, the decentralized nature of blockchain ensures that the compromised device cannot single-handedly manipulate or compromise the entire system. The distributed ledger technology also makes it more challenging for attackers to manipulate or delete data (Bhambri & Rani, 2024).

By combining edge computing and blockchain, IoT systems can benefit from enhanced privacy, data integrity, resilience, and efficient device management, ultimately improving overall security in the rapidly expanding world of interconnected devices.

11.7 POTENTIAL CHALLENGES AND LIMITATIONS OF THE COMBINED APPROACH

The combined approach of edge computing and blockchain technology in the IoT context offers numerous benefits but also faces some challenges and limitations. Scalability becomes a concern as edge computing involves managing processing power and storage across distributed edge devices, while blockchain requires nodes to validate and store transactions. As the number of IoT devices and transactions increases, scalability becomes more challenging. Preserving privacy in IoT systems is another challenge. Recently, numerous privacy and security flaws have been identified for emerging technologies, including the vulnerabilities of cryptography-based key management schemes against quantum computers (Chorti et al., 2022), and the misuse of massive private data by service providers (Falchuk et al., 2018).

While edge computing aims to reduce latency by processing data closer to the source, integrating blockchain introduces additional delays. Blockchain consensus algorithms, like Proof of Work or Proof of Stake, require time-consuming computations that can impact real-time responsiveness in certain IoT applications. Furthermore, blockchain networks, particularly those using Proof of Work, consume significant computational power and energy, posing challenges for resource-constrained edge devices with limited processing capabilities and battery life. Ensuring interoperability between edge computing and blockchain technologies is complex due to the entanglement of diverse devices and platforms in IoT systems. Developing standardized interfaces and protocols is essential for seamless integration. Deploying edge computing and blockchain solutions incurs upfront costs and requires specialized expertise.

Addressing these challenges necessitates ongoing research, innovation, and collaboration among industry stakeholders. Overcoming these limitations will contribute to the wider adoption and successful implementation of edge computing and blockchain in IoT applications.

11.8 FUTURE DIRECTIONS

The emerging opportunities, possible future trends, and technologies in IoT security can be developed for industry initiatives with standardization efforts. In the dynamic world of IoT technology, security is a critical concern. To address emerging challenges and protect IoT devices, networks, and data, several trends and technologies are gaining traction in IoT security. These include device identity and authentication, secure communication protocols, integration of 5G networking, blockchain for IoT security, edge computing, fog computing, security analytics and machine learning, regulatory frameworks, and standards. Authentication and authorization mechanisms are being strengthened through multi-factor authentication, biometrics, and public key Infrastructure (PKI) to verify device identity. Edge computing is gaining prominence as it enables data processing closer to the source, reducing latency and enhancing security by minimizing data transmission to the cloud. Blockchain technology is being explored to ensure trust, transparency, and data integrity in IoT transactions. AI and ML technologies are increasingly utilized to detect anomalies and respond to security incidents in real time. Security-by-design is being embraced, integrating security features into IoT devices from the design phase. Enhanced data encryption techniques and protocols are being developed to protect data in transit and at rest. Regulatory frameworks are being established to enforce security measures and privacy requirements for IoT deployments. Security analytics and threat intelligence platforms are being leveraged to proactively identify vulnerabilities and respond effectively to potential attacks. As the IoT security landscape continues to evolve, these trends are likely to shape the industry's future direction. It is important for organizations to stay updated on these developments and implement robust security measures to safeguard their IoT ecosystems.

The field of IoT security offers several promising future opportunities as the proliferation of connected devices continues. These opportunities include developing advanced threat detection and response mechanisms using AI and ML, automating security processes across diverse IoT environments, ensuring privacy and data protection, and establishing secure firmware update and patch management systems. Lone et al. (2023) provide a comprehensive overview of the cybersecurity challenges and opportunities in the IoT domain. The authors also highlight the architectural layers of the IoT and the possible attacks against these IoT layers. These opportunities arise from the increasing complexity and importance of securing IoT ecosystems, providing room for innovation and growth in the industry. Staying updated on these advancements is crucial as IoT security continues to evolve rapidly.

11.9 CONCLUSION

In summary, the fusion of edge computing with blockchain offers a viable way to improve IoT security. The complex security difficulties of IoT systems can be tackled by combining the strengths of blockchain, such as decentralized trust and tamper-resistant data storage, with those of edge computing, such as real-time data processing and localized security procedures. This chapter intends to encourage additional innovation and promote breakthroughs in the field of IoT security, enabling the secure deployment of IoT technologies through an analysis of existing research and solutions. This chapter intends to encourage additional innovation and promote development in the area of IoT security, with the ultimate goal of facilitating the safe implementation of IoT solutions.

11.10 DATA AVAILABILITY STATEMENT

Data-sharing is not applicable to this chapter as no new data were created or analyzed in this study.

REFERENCES

Alam, M., & Siddiqui, M. I. (2023). Effective framework to tackle urban unemployment by e-government: An IoT solution for smart/metro cities in developing nation. *Journal of Science and Technology Policy Management*, *14*(1), 213–238.

Bali, V., Bali, S., Gaur, D., Rani, S., & Kumar, R. (2023). Commercial-off-the shelf vendor selection: A multi-criteria decision-making approach using intuitionistic fuzzy sets and TOPSIS. *Operational Research in Engineering Sciences: Theory and Applications*, *6*(2).

Bhambri, P., & Rani, S. (2024). Challenges, opportunities, and the future of industrial engineering with IoT and AI. *Integration of AI-Based Manufacturing and Industrial Engineering Systems with the Internet of Things*, 1–18.

Bhambri, P., Rani, S., Balas, V. E., & Elngar, A. A. (Eds.). (2023). *Integration of AI-Based Manufacturing and Industrial Engineering Systems with the Internet of Things*. USA: CRC Press.

Caparra, G., Centenaro, M., Laurenti, N., Tomasin, S., & Vangelista, L. (2017). 14 wireless physical-layer authentication for the internet of things. *Information Theoretic Security and Privacy of Information Systems*, *390*.

Chauhan, M., & Rani, S. (2021). Covid-19: A revolution in the field of education in India. *Learning How to Learn Using Multimedia*, 23–42.

Chen, D., Zhang, N., Cheng, N., Zhang, K., Qin, Z., & Shen, X. (2018). Physical layer based message authentication with secure channel codes. *IEEE Transactions on Dependable and Secure Computing*, *17*(5), 1079–1093.

Chen, S., Li, Q., Zhou, M., & Abusorrah, A. (2021). Recent advances in collaborative scheduling of computing tasks in an edge computing paradigm. *Sensors*, *21*(3), 779.

Chorti, A., Barreto, A. N., Köpsell, S., Zoli, M., Chafii, M., Sehier, P., . . . Poor, H. V. (2022). Context-aware security for 6G wireless: The role of physical layer security. *IEEE Communications Standards Magazine*, *6*(1), 102–108.

Dedeoglu, V., Jurdak, R., Putra, G. D., Dorri, A., & Kanhere, S. S. (2019, November). A trust architecture for blockchain in IoT. In *Proceedings of the 16th EAI International Conference on Mobile and Ubiquitous Systems: Computing, Networking and Services* (pp. 190–199). UK.

Deepa, N., Pham, Q. V., Nguyen, D. C., Bhattacharya, S., Prabadevi, B., Gadekallu, T. R., . . . Pathirana, P. N. (2022). A survey on blockchain for big data: Approaches, opportunities, and future directions. *Future Generation Computer Systems*, *131*, 209–226.

Falchuk, B., Loeb, S., & Neff, R. (2018). The social metaverse: Battle for privacy. *IEEE Technology and Society Magazine*, *37*(2), 52–61.

Fati, S. M., Muneer, A., Mungur, D., & Badawi, A. (2018, July). Integrated health monitoring system using GSM and IoT. In *2018 International Conference on Smart Computing and Electronic Enterprise (ICSCEE)* (pp. 1–7). New York, NY: IEEE.

Guo, H., & Yu, X. (2022). A survey on blockchain technology and its security. *Blockchain: Research and Applications*, *3*(2), 100067.

Haddad Pajouh, H., Dehghantanha, A., Khayami, R., & Choo, K. K. R. (2018). A deep recurrent neural network based approach for internet of things malware threat hunting. *Future Generation Computer Systems*, *85*, 88–96.

Kataria, A., Agrawal, D., Rani, S., Karar, V., & Chauhan, M. (2022). Prediction of blood screening parameters for preliminary analysis using neural networks. In *Predictive Modeling in Biomedical Data Mining and Analysis* (pp. 157–169). New York, NY: Academic Press.

Kataria, A., Puri, V., Pareek, P. K., & Rani, S. (2023, July). Human activity classification using G-XGB. In *2023 International Conference on Data Science and Network Security (ICDSNS)* (pp. 1–5). Raichur: IEEE.

Kouicem, D. E., Bouabdallah, A., & Lakhlef, H. (2018). Internet of things security: A top-down survey. *Computer Networks*, *141*, 199–221.

Kumar, P., Banerjee, K., Singhal, N., Kumar, A., Rani, S., Kumar, R., & Lavinia, C. A. (2022). Verifiable, secure mobile agent migration in healthcare systems using a polynomial-based threshold secret sharing scheme with a blowfish algorithm. *Sensors*, *22*(22), 8620.

Lone, A. N., Mustajab, S., & Alam, M. (2023). A comprehensive study on cybersecurity challenges and opportunities in the IoT world. *Security and Privacy*, *6*(6), e318.

Luo, Y., Su, Z., Zheng, W., Chen, Z., Wang, F., Zhang, Z., & Chen, J. (2021). A novel memory-hard password hashing scheme for blockchain-based cyber-physical systems. *ACM Transactions on Internet Technology (TOIT)*, *21*(2), 1–21.

Majeed, U., Khan, L. U., Yaqoob, I., Kazmi, S. A., Salah, K., & Hong, C. S. (2021). Blockchain for IoT-based smart cities: Recent advances, requirements, and future challenges. *Journal of Network and Computer Applications*, *181*, 103007.

Maughelli, F. (2016). *Security-Related Experiences with Smart Contracts Over the Ethereum Blockchain*. https://amslaurea.unibo.it/13991/1/francesco_maughelli_tesi.pdf.

Miglani, A., Kumar, N., Chamola, V., & Zeadally, S. (2020). Blockchain for internet of energy management: Review, solutions, and challenges. *Computer Communications*, *151*, 395–418.

Mijuskovic, A., Chiumento, A., Bemthuis, R., Aldea, A., & Havinga, P. (2021). Resource management techniques for cloud/fog and edge computing: An evaluation framework and classification. *Sensors*, *21*(5), 1832.

Puri, V., Kataria, A., Solanki, V. K., & Rani, S. (2022, December). AI-based botnet attack classification and detection in IoT devices. In *2022 IEEE International Conference on Machine Learning and Applied Network Technologies (ICMLANT)* (pp. 1–5). Salvador: IEEE.

Puthal, D., Nepal, S., Ranjan, R., & Chen, J. (2016). Threats to networking cloud and edge datacenters in the internet of things. *IEEE Cloud Computing*, *3*(3), 64–71.

Qadri, I., Muneer, A., & Fati, S. M. (2021, February). Automatic robotic scanning and inspection mechanism for mines using IoT. In *IOP Conference Series: Materials Science and Engineering* (Vol. 1045, No. 1, p. 012001). USA: IOP Publishing.

Rani, S., Bhambri, P., & Kataria, A. (2023). Integration of IoT, big data, and cloud computing technologies. In *Big Data, Cloud Computing and IoT: Tools and Applications*. USA: CRC Press.

Rani, S., Kataria, A., & Chauhan, M. (2022). Cyber security techniques, architectures, and design. In *Holistic Approach to Quantum Cryptography in Cyber Security* (pp. 41–66). USA: CRC Press.

Rani, S., Kataria, A., Kumar, S., & Tiwari, P. (2023). Federated learning for secure IoMT-applications in smart healthcare systems: A comprehensive review. *Knowledge-Based Systems*, 110658.

Rani, S., Kaur, J., & Bhambri, P. (2023). Technology and gender violence: Victimization model, consequences and measures. In *Communication Technology and Gender Violence* (pp. 1–19). Cham: Springer International Publishing.

Rani, S., Kumar, S., Kataria, A., & Min, H. (2023). SmartHealth: An intelligent framework to secure IoMT service applications using machine learning. *ICT Express*, *10*(2), 420–425.

Rani, S., Mishra, A. K., Kataria, A., Mallik, S., & Qin, H. (2023). Machine learning-based optimal crop selection system in smart agriculture. *Scientific Reports*, *13*(1), 15997.

Rani, S., Pareek, P. K., Kaur, J., Chauhan, M., & Bhambri, P. (2023, February). Quantum machine learning in healthcare: Developments and challenges. In *2023 IEEE International Conference on Integrated Circuits and Communication Systems (ICICACS)* (pp. 1–7). Raichur: IEEE.

Satyanarayanan, M., Simoens, P., Xiao, Y., Pillai, P., Chen, Z., Ha, K., . . . Amos, B. (2015). Edge analytics in the internet of things. *IEEE Pervasive Computing*, *14*(2), 24–31.

Singh, H., & Rani, S. (2023). Flex sensor integrated smart strap to verify correct wearing of the face mask. *IEEE Sensors Journal*, *24*(2), 2020–2027.

Sudevan, S., Barwani, B., Al Maani, E., Rani, S., & Sivaraman, A. K. (2021). Impact of blended learning during Covid-19 in Sultanate of Oman. *Annals of the Romanian Society for Cell Biology*, 14978–14987.

Tanwar, R., Chhabra, Y., Rattan, P., & Rani, S. (2022, September). Blockchain in IoT networks for precision agriculture. In *International Conference on Innovative Computing and Communications: Proceedings of ICICC 2022* (Vol. 2, pp. 137–147). Singapore: Springer Nature Singapore.

Wu, Y., Dai, H. N., & Wang, H. (2020). Convergence of blockchain and edge computing for secure and scalable IoT critical infrastructures in industry 4.0. *IEEE Internet of Things Journal*, *8*(4), 2300–2317.

Xiao, Y., Jia, Y., Liu, C., Cheng, X., Yu, J., & Lv, W. (2019). Edge computing security: State of the art and challenges. *Proceedings of the IEEE*, *107*(8), 1608–1631.

Xu, C., Qu, Y., Luan, T. H., Eklund, P. W., Xiang, Y., & Gao, L. (2021). A lightweight and attack-proof bidirectional blockchain paradigm for Internet of Things. *IEEE Internet of Things Journal*, *9*(6), 4371–4384.

Zhang, L., Zhang, Z., Wang, W., Jin, Z., Su, Y., & Chen, H. (2021). Research on a covert communication model realized by using smart contracts in blockchain environment. *IEEE Systems Journal*, *16*(2), 2822–2833.

12 Future Prospects of Blockchain for Secure IoT Systems

Vikram Puri, Aman Kataria, Sita Rani, and Chandra Shekar Pant

12.1 INTRODUCTION

Due to frequent cyberattacks that disrupt and weaken networks, security is a worrying challenge in the networking system. The extensive integration of sensors and cameras heightens the network security issue. Even though businesses use a variety of strategies to address this problem, including network access control, information encryption, firewalls, and real-time vulnerability, data is still modified in networks by third parties. The following is an introduction to and discussion of the concept of autonomous systems for the convenience of people's lifestyles.

12.1.1 Internet of Things (IoT)

IoT is an interconnected collection of everyday things with built-in real-time sensors, software, and network access to gather and share data (Rani et al., 2023). It enables connections between these devices, information sharing, and control of the devices from outside the network. Increasing the effectiveness of industrial processes, recycling waste, and automating manufacturing operations have revolutionized several industries in various ways (Khanna & Kaur, 2020). Additionally, it improves decision-making ability and the customer experience by identifying risks. The demand for IoT devices is rising, which attracts hackers and leaves networks open to attack. IoT security protects devices inside and outside of the network against malware, unauthorized access, and data encryption (Puri et al., 2022). The sheer volume of devices that are hard to track, the difficulty distinguishing between real and fake devices, and the difficulty identifying devices that increase system vulnerability are some significant difficulties facing the IoT network. The application of IoT in the real world is limitless and has now extended to various domains. The most popular IoT application areas are as follows:

a. *Smart Cities:* IoT is essential to creating smart cities since it facilitates the data collection process and automates routine systems like waste management, transportation, logistics, and traffic management (Rani et al., 2022).

 DOI: 10.1201/9781003460367-12

b. *Healthcare Ecosystem:* IoT has the potential to improve the healthcare environment (Singh et al., 2022), including surgical procedures and remote patient monitoring. It also makes it possible to provide tailored care, manage patient health records, and give doctors access to virtual assistants to help them make better decisions.
c. *Agriculture:* IoT has completely changed the agriculture sector by increasing crop production in both quality and quantity, cutting waste, and boosting system performance. IoT is mostly used in agriculture (Farooq et al., 2022) to track soil nutrients, which enables farmers to identify crop illnesses early on and automate farming procedures and supply chain management.
d. *Internet of Industrial Things (IIoT) and Energy Management:* IoT is crucial to the IIoT industry's efforts to lower costs and increase system efficiency. Real-time monitoring and smart grid management are primary goals of IoT to improve efficiency and increase industry profits (Peter et al., 2023). Additionally, IoT devices can monitor device energy usage in real time and identify patterns for energy consumption that can immediately result in energy savings.
e. *Logistics and Transportation:* IoT enables businesses to streamline operations, cut costs, and improve customer experience in logistics and transportation (Kumar et al., 2022). It has several important roles to play in logistics and transportation, including real-time shipment tracking, providing real-time traffic information to optimize shipment routing and save fuel, and monitoring the vehicle's environmental conditions to ensure that goods are transported in proper environmental conditions, especially for medicines.
f. *Wearables:* These days, trackers and wearable technology are increasingly commonplace. IoT devices enhance device usefulness in a variety of ways, including data collecting and analysis, real-time vital signs monitoring, smart decision-making based on parameters and vitals, and most crucially, the ability to interact with other smart devices (Baldini et al., 2023).
g. *Autonomous Vehicles:* Connected self-driving cars communicate data from onboard sensors, devices belonging to cyclists and walkers, traffic cameras, parking detection devices, etc. The IoT connectivity processes the data from various sensors, draws a route, and delivers commands to the brake, steering, and accelerate systems of the vehicle (Shah et al., 2022). IoT makes it feasible to gather the massive quantity of data that these autonomous vehicles require from the road, sensors, and obstacles in front of the automobile.

These applications are all currently relying on centralized systems like cloud computing. Even though these centralized systems are generally effective, they have certain drawbacks, including security of data risks, energy use, single points of failure, and managing enormous amounts of data.

12.1.2 Blockchain Technology

Blockchain technology is a decentralized network that holds transaction logs, or blocks, in a network linked by nodes communicating. The blockchain network is more reliable and secure because the data is immutable or cannot be removed or changed (Rajasekaran et al., 2022). Blockchain technology is typically best recognized for the cryptocurrency system for safely and openly tracking transactions. However, blockchain technology is not just for cryptocurrencies and has applicable industrial uses. The following are the merits to introducing blockchain technology in various industries:

a. *Decentralization:* Blockchain operates on a decentralized network; thus, no single entity is in charge of the data. Decentralization also makes it possible to improve system transparency and address single points of failure.
b. *Immutability:* A further advantage of blockchain technology is that once data is registered on the network, it cannot be changed or tempered without consensus. This boosts the blockchain network's security.
c. *Security:* The blockchain network uses cryptography to secure transactions and data. The use of cryptography raises significant barriers to access for outsiders or hackers.
d. *Transparency:* A transparent and verifiable ledger of transactions is provided by blockchain technology. Since each exchange is captured on the blockchain and can be tracked back to its source, accountability is increased and fraud is decreased.
e. *Efficient and Cost-Saving:* Blockchain replaces the central access point, automates digital transactions, and saves time, money, and paperwork compared to the traditional approach.
f. *Trust and Collaboration:* It facilitates collaboration and cooperation between two entities with transparent and immutable information, which can foster confidence between them, thanks to the immutable record of transaction.

The limitations of centralized or traditional platforms can be solved by blockchain technology. IoT technology is now a crucial area for humankind, but difficulties and failures make it challenging to deploy in the real world. Blockchain technology can resolve these problems and restore humanity's trust in itself. Figure 12.1 represents the challenges of IoT in different applications.

12.2 SYNERGY OF BLOCKCHAIN TECHNOLOGY WITH IoT ECOSYSTEM

IoT systems are vulnerable to a wide range of security risks and threats, including lack of infrastructure hardening, which means that most devices are deployed in public or remote locations where they can be accessed by hackers and tamper with the data; the overwhelming majority of IoT devices not adhering to proper encryption to protect the sensitive information which can be leaked, vulnerable web applications and APIs, which can also result in the system being compromised. Similar IoT devices are being used as botnets to attack other systems.

Numerous studies have reported using encrypted sensitive information (Hans et al., 2022), robust access and authenticate control management systems (Trnka

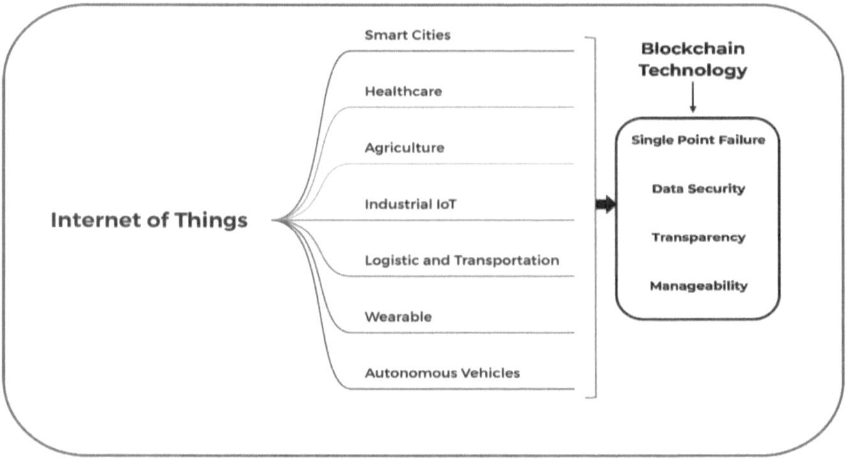

FIGURE 12.1 Challenges of IoT in different applications.

et al., 2022), and other techniques to address these problems. However, the system can be hacked or altered, and this technique only solves the problem temporarily. The following methods can be used to secure IoT using blockchain:

A. *Secure:* When storing data, blockchain adds an extra layer of protection that makes it challenging for hackers or any other third party to go around and access this critical information inside the network. Furthermore, blockchain technology enables machine-to-machine transactions, which can improve node tracing and data security. Moin et al. (2019) presented a review study on how blockchain technology changed the IoT ecosystem and the underlying problems for integrating blockchain technology into the IoT system. Additionally, it offered a taxonomy of advantages, disadvantages, and chances for improvement.
B. *Lower Costing:* By eliminating third-party verification, which frequently entails fees, the overhead cost can be decreased by incorporating blockchain technology. Additionally, businesses can streamline their payment procedures and even do away with or minimize transaction fees, thanks to blockchain technology. In Kfoury (2021), a study regarding how blockchain technology can reduce financial transactions in a different method is presented, along with a study based on related studies and expert interviews. Similarly, eliminating intermediaries or centralized authorities in an IoT system (Da Xu et al., 2021) can reduce system costs for transactions and services.
C. *Transparency and Tamper-Proof Data:* In general, blockchain technology logs every transaction in a digital ledger that contains a secure, unalterable data chain. Blockchain technology enables IoT devices to store and communicate data, preserving its truthfulness and building participant confidence safely and openly. By reducing data modification and unwanted access, this

openness also improves the IoT system's safety. IoT devices are hard to hack or manipulate and introduce botnets into the network since they have transparency and tamper prevention. Some recent studies have focused on the digital signature algorithms that create barriers for hackers or third-party entities. In Janani & Ramamoorthy (2023), there is proof of IoT designed for the IoT ecosystem to counter cyberattacks during the data communication between the devices.

D. *Enhanced Privacy:* IoT devices must be developed with privacy considerations from the outset because they gather extensive private information, are prone to data theft and hacking breaches, need visibility and approval, and require these features. Blockchain offers a practical remedy for IoT system privacy. It establishes a virtual ledge where all transactions involving IoT devices are kept. These transactions include timestamps, data acquired, and meta-information devices meta information. However, the identities of the gadgets are kept secret to preserve privacy. The blockchain-based distributed entity is proposed (Yin et al., 2022) to provide strong privacy for IoT privacy.

E. *Interoperability:* Multiple options exist for blockchain technology to enhance IoT device interoperability. First, blockchain offers a decentralized framework that facilitates system and component interoperability. With no need for a centralized authority or intermediary, gadgets may now converse and exchange data with one another. Second, a secure, transparent ledger that keeps track of every transaction and information transfer is another way that blockchain may guarantee the safety and authenticity of IoT data. Doing so can guard against information theft and guarantee the reliability and accuracy of your data. Third, smart contracts are self-executing agreements that can be set up to take effect when specific criteria are met. They can be utilized to simplify and automate IoT procedures, including data transfer, device verification, and handling payments. The "blockchain of blockchain" is a hierarchical structure of blockchains for intelligent cities introduced by Rahman et al. (2022) to guarantee data accuracy and interoperability simultaneously.

F. *Self-Automation:* IoT devices with smart contracts and the blockchain have much potential because they allow for self-automation and scalable authentication. In the blockchain network, smart contracts enable automated processes and offer safe authentication controls. A proposed study (Hewa et al., 2021) is to determine the key technological components of blockchain-based smart contracts and the related future research objectives.

G. *Efficient Supply Chain Management:* The primary gains are improved traceability, cost and waste reduction, increased security through cryptographic techniques, and stakeholder access to real-time data without worrying about their privacy and trust in the system. This study (Rejeb et al., 2019) suggested how blockchain technology and the IoT system may work together to improve network performance and streamline supply chain management. Additionally, it examines the integration approach's research proposition on essential characteristics, including reliability, safety, indestructibility,

TABLE 12.1
Studies Based on Integration of IoT and Blockchain System

Study	Decentralization	Costing	Privacy	Decentralized Storage
(Moin et al., 2019)	✓	×	×	×
(Kfoury, 2021)	✓	✓	×	×
(Da Xu et al., 2021)	✓	✓	×	×
(Janani & Ramamoorthy, 2023)	✓	×	×	×
(Yin et al., 2022)	✓	×	✓	×
(Rahman et al., 2022)	✓	×	×	×
(Hewa et al., 2021)	✓	×	×	×
(Rejeb et al., 2019)	✓	×	✓	✓

traceability, quality, and interoperability. It goes on to explain how these characteristics may affect future methods of doing things.

H. *Decentralization:* In the IoT ecosystem, the blockchain system allows control and decision-making from the centralized system to the distributed network. Decentralized networks seek to limit the amount of trust that members must place in one another and to prevent them from interfering with one another's abilities to exercise authority or control in ways that harm the network's performance. The hash of the encrypted data created by IoT devices is saved in the blockchain to enable decentralized administration of that data, and the raw data is subsequently kept in a safe platform using a trusted execution ecosystem. Decentralized virtual payment is suggested in this study (Sahoo & Chaurasiya, 2023). With the help of smart contracts and Ethereum, system performance may be measured. Additionally, IPFS and the SHA3 method are used for data storage. Table 12.1 presents the studies based on integrating IoT and blockchain systems.

12.3 THE FUTURE OF BLOCKCHAIN FOR SECURE IoT SYSTEMS?

The worldwide blockchain IoT market's future is bright, and industry-wide transformational effects are predicted. The following sections explore how blockchain can be used to protect various IoT systems (see Figure 12.2):

a. *Smart Cities:* To enable immediate data exchange among residents, authorities, and enterprises, smart cities, an emerging idea, incorporate technology for communication and information. A smart city's numerous stakeholders may communicate and transact with one another with a new degree of convenience and security, thanks to the combination of IoT and blockchain technologies (Majeed et al., 2021). The significant uses of the IoT and blockchain in the smart cities are as follows:

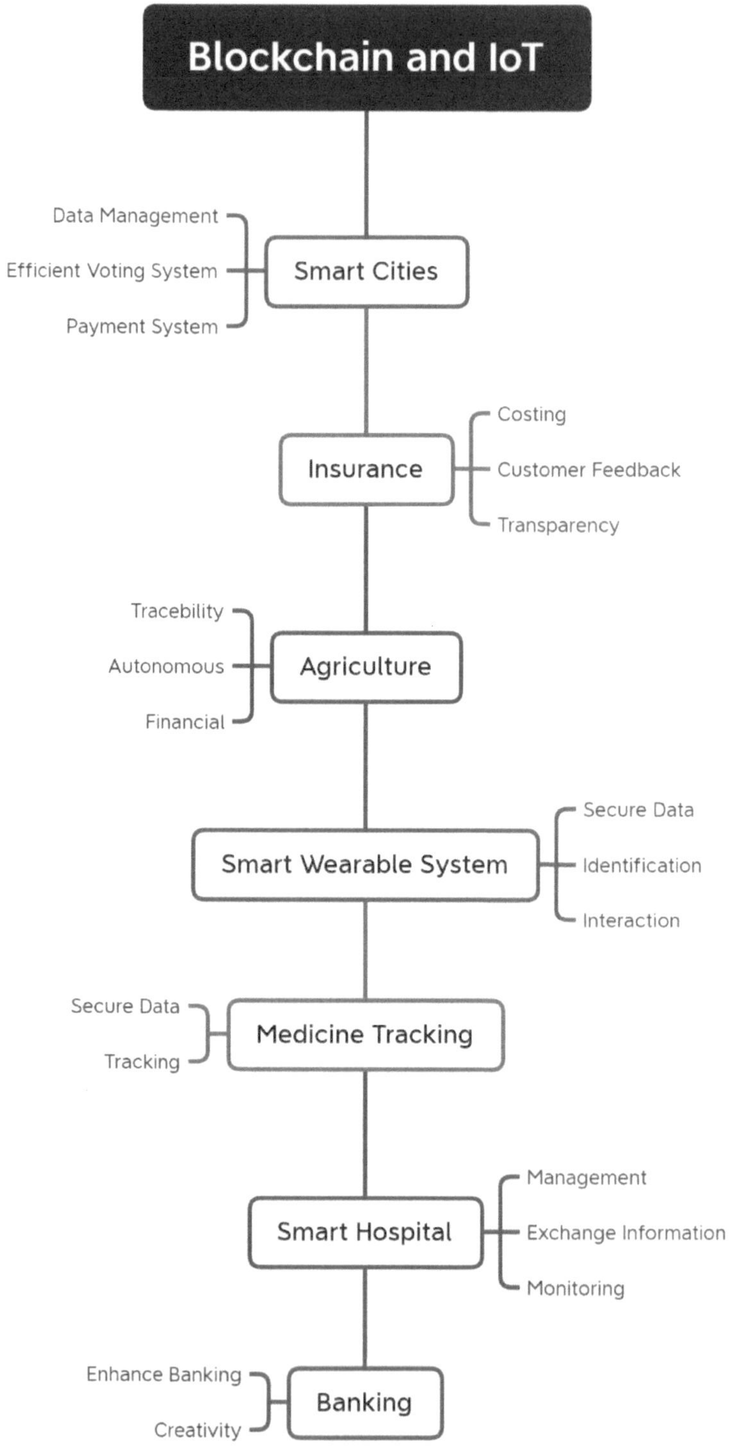

FIGURE 12.2 Integration approach of blockchain and IoT for different sectors.

- *Data Management:* IoT gadgets can gather instantaneous information on several smart city components, including traffic, energy use, and pollution levels. A distributed ledger called blockchain can be used to securely store this information securely, thereby ensuring its accuracy, confidentiality, and visibility.
- *Efficient Voting System:* Thanks to blockchain technology, voting systems that are fair and safe can let people engage in elections. IoT devices can gather information about citizen preferences and actions, which can be saved on a decentralized network for safe and open decision-making.
- *Payment System:* Blockchain technology can automate transactions in intelligent cities, cutting transaction costs and eliminating intermediaries. For safe and efficient payments, connected devices can gather information on transactions involving payments, which can then be stored on a blockchain.

b. *Insurance:* With various advantages and prospects for both insurers and policyholders, blockchain technology is significantly changing the insurance sector (Xiao et al., 2020). The following are discussing about the role of blockchain and IoT in the insurance sector:
    - *Costing:* The financial services industry can save money by streamlining and automating numerous operations with the use of blockchain technology. For instance, it can streamline the claims registration procedure, removing the need for paper-based paperwork and lowering administrative expenses.
    - *Customer Feedback:* Insurance companies might provide clients with more confidence and transparency by utilizing blockchain. Because blockchain is decentralized, it enables safe, unchangeable records, guaranteeing that policyholders have access to reliable data that cannot be altered.
    - *Transparency:* By allowing all participants to an insurer exchange to have ownership of identical data, the use of distributed ledgers on the blockchain fosters openness and confidence. Due to the easier detection and prevention of forged documents, this can aid in preventing fraud as well as abuse.

c. *Agriculture:* IoT and blockchain can fundamentally alter the farming industry (Bhat et al., 2021). There are some numerous ways to implement that:
    - *Traceability and Management:* Using blockchain technology, the food supply chain may be transparent, and data can be traced, increasing food safety. IoT data and blockchain can be combined to trace and authenticate farm product sources, quality, and management.
    - *Autonomous:* The blockchain system can be used to construct smart contracts, which are autonomous agreements with established rules, that can streamline numerous agricultural activities. Utilizing real-time information, IoT devices can start events in the blockchain system, allowing automated operations and making choices.
    - *Financial:* By eradicating middlemen and enabling the formation of decentralized marketplaces, blockchain technology can link farmers and customers directly. Additionally, it may make it possible for

farmers to employ agricultural digital currencies, increasing their financial standing and giving them the ability to use banking services.

d. *Smart Wearable System:* Blockchain and IoT unlock the potential of smart wearable devices (Puri et al., 2021). There are some following ways:
   - *Secure Data:* Sensitive health data can be collected via mobile devices, and blockchain technology may assist in encrypting this data by building a decentralized, invulnerable to record of interactions.
   - *Identification:* Smart wristbands can maintain identities securely and autonomously by using blockchain technology, giving consumers full control over their identity and data. This can improve transaction efficiency and security while also enhancing privacy.
   - *Interaction:* Smart devices using blockchain technology may be able to provide users with the option of earning rewards for disclosing their personal information to third parties. This may promote greater data-sharing and open new avenues for scientific investigation and advancement.

e. *Medicine Tracking:* For handling the tracking of medications in the pharmaceutical supply chain, blockchain and IoT can be associated. Both can work together to track medical supplies securely and effectively.
   - *Secure Data:* The use of blockchain-based technologies can enhance IoT and healthcare data management processes. Blockchain's autonomous nature fosters digital environment confidence, which is crucial for the medical distribution system.
   - *Tracking:* The distribution network for medications can be tracked and traced using a system based on blockchain powered by IoT sensors. This may assist in ensuring that the medication is carried and stored at the proper temperature as well as preventing the market entry of counterfeit pharmaceuticals.

f. *Smart Hospital:* Smart hospitals can employ blockchain and IoT to enhance monitoring of patients, communication between devices, and data administration integrity (Sharma et al., 2021).
   - *Management:* Critical medical information may be shared and stored with an elevated degree of confidentiality due to blockchain technology. It can overcome problems with privacy, security of data, collaborating, and holding to guarantee that health data are protected in the healthcare system with the highest level of openness.
   - *Exchanging Information:* The medical condition of patients can be recorded by IoT devices, processed, and sent to the ledger. This makes it possible for equipment, healthcare facilities, and medical professionals to share data. Additionally, blockchain can be used to manage the drug supply chain, exchange patient medical records securely, and assist with genetic study in the field of healthcare.
   - *Monitoring:* Smart medical devices built on the IoT can collect important data and offer more perspectives on conditions and actions. Blockchain can allow patients to share the real-time data generated by IoT devices with the doctor once they have logged into the application.

g. *Banking:* Banks may develop cutting-edge solutions that enhance security, effectiveness, and customer satisfaction by fusing blockchain and IoT (Surekha et al., 2022). For instance, blockchain can offer an open and trustworthy framework for IoT devices for sharing data, protecting the confidentiality and integrity of the data.
   - *Enhance Security:* Banks may provide a safe and open environment for money transfers by fusing blockchain with IoT. Blockchain offers a safe and impenetrable method for storing and verifying transactions, and IoT devices can add more layers of verification, such as fingerprint sensors, to boost security.
   - *Creativity:* Banks are able to develop cutting-edge solutions that enhance customer service by fusing blockchain and IoT. By gathering information on customer needs and actions, IoT devices, for instance, can offer tailored banking occasions, while blockchain technology can be used to create smart contracts that automate different banking procedures.

12.4 CHALLENGES AND ISSUES OF BLOCKCHAIN FOR THE SECURE IoT SYSTEM

Even though using blockchain in IoT introduces several challenges and issues, it can assist in addressing some of the security issues that IoT networks confront. These issues are as follows:

- *Sensors Reliability:* IoT devices employ sensors to collect information, and if they are faulty, this may lead to inaccurate information and security issues. In terms of IoT data security, sensor dependability is crucial. To ensure data integrity, sensor robustness and accuracy must be validated. To maintain optimal operation, sensors should also be frequently checked and updated.
- *Privacy:* Blockchain technology necessitates a decentralized network, which might compromise privacy if managed incorrectly. Due to blockchain technology's public and decentralized nature, all data is kept on a distributed ledger that is open to all users. Identity theft, data misuse, and harmful assaults are more likely as a result. Therefore, it is crucial to make sure blockchain technology is maintained appropriately.
- *Computation Power:* Due to their processing complexity, substantial communication overhead, and latency, several IoT devices cannot leverage blockchain technology. However, blockchain technology can be used in some use scenarios where IoT devices need safe information storage, decentralized authentication, and impermeable transactions. Blockchain technology, for instance, can allow IoT gadgets to communicate data, safely keep information, and ensure data validity.
- *Centralization Impact:* As the capacity of the blockchain's globally accessible database expands gradually, authoritarianism may result from each entity having access to the entire record and its complete history.

- *Hardware:* It can be challenging to link the equipment of numerous IoT systems with a public decentralized blockchain because they have comparatively little RAM and storage capacity.
- *Scalability:* The high computation and storage needs of blockchain technology can make employing many IoT devices with limited resources challenging. While maintaining the required security, other protocols like Hyperledger or Ethereum can leverage off-chain processing to lower the required resources. These protocols are feasible for IoT devices because they offer scalability and privacy.

12.5 CONCLUSION

Despite the challenges, the amalgamation of blockchain and IoT has much potential. By providing decentralized and tamper-proof data storage and communication, merging these two cutting-edge technologies can improve security. Blockchain technology can offer a safe and decentralized platform for IoT devices to interact and exchange data. IoT sensors and blockchain technology are anticipated to be widely used by medical and government agencies to improve the safety of healthcare facilities, buildings, day-to-day lifestyles and communities during upcoming epidemics. In this chapter, blockchain-enabled IoT systems and the benefits of blockchain are discussed. The potential applications of blockchain in various IoT network sectors have been examined, and the future of blockchain to secure the IoT ecosystem has been highlighted.

Integration of IoT and blockchain demonstrates a high degree of confidence, and the prospect of blockchain participation in the IoT ecosystem is encouraging. This is because it can offer a safe, open, and decentralized platform for information exchange and handling transactions. Moreover, blockchain is a perfect option for deployment in the IoT environment because of its capacity to keep data safely and reliably.

REFERENCES

Baldini, A., Garofalo, R., Scilingo, E. P., & Greco, A. (2023). A real-time, open-source, IoT-like, wearable monitoring platform. *Electronics*, 12(6), 1498.

Bhat, S. A., Huang, N. F., Sofi, I. B., & Sultan, M. (2021). Agriculture-food supply chain management based on blockchain and IoT: A narrative on enterprise blockchain interoperability. *Agriculture*, 12(1), 40.

Da Xu, L., Lu, Y., & Li, L. (2021). Embedding blockchain technology into IoT for security: A survey. *IEEE Internet of Things Journal*, 8(13), 10452–10473.

Farooq, M. S., Sohail, O. O., Abid, A., & Rasheed, S. (2022). A survey on the role of IoT in agriculture for the implementation of smart livestock environment. *IEEE Access*, 10, 9483–9505.

Hans, S., Ghosh, S., Kataria, A., Karar, V., & Sharma, S. (2022). Controller placement in software defined internet of things using optimization algorithm. *Computers, Materials & Continua*, 70(3).

Hewa, T. M., Hu, Y., Liyanage, M., Kanhare, S. S., & Ylianttila, M. (2021). Survey on blockchain-based smart contracts: Technical aspects and future research. *IEEE Access*, 9, 87643–87662.

Janani, K., & Ramamoorthy, S. (2023). A security framework to enhance IoT device identity and data access through blockchain consensus model. *Cluster Computing*, 1–24.

Kfoury, B. (2021). The role of blockchain in reducing the cost of financial transactions in the retail industry. In *CEUR Workshop Proceedings* (Vol. 2889, pp. 10–22). UK.

Khanna, A., & Kaur, S. (2020). Internet of things (IoT), applications and challenges: A comprehensive review. *Wireless Personal Communications*, 114, 1687–1762.

Kumar, D., Singh, R. K., Mishra, R., & Wamba, S. F. (2022). Applications of the internet of things for optimizing warehousing and logistics operations: A systematic literature review and future research directions. *Computers & Industrial Engineering*, 108455.

Majeed, U., Khan, L. U., Yaqoob, I., Kazmi, S. A., Salah, K., & Hong, C. S. (2021). Blockchain for IoT-based smart cities: Recent advances, requirements, and future challenges. *Journal of Network and Computer Applications*, 181, 103007.

Moin, S., Karim, A., Safdar, Z., Safdar, K., Ahmed, E., & Imran, M. (2019). Securing IoTs in distributed blockchain: Analysis, requirements and open issues. *Future Generation Computer Systems*, 100, 325–343.

Peter, O., Pradhan, A., & Mbohwa, C. (2023). Industrial internet of things (IIoT): Opportunities, challenges, and requirements in manufacturing businesses in emerging economies. *Procedia Computer Science*, 217, 856–865.

Puri, V., Kataria, A., & Sharma, V. (2021). Artificial intelligence-powered decentralized framework for internet of things in healthcare 4.0. *Transactions on Emerging Telecommunications Technologies*, e4245.

Puri, V., Kataria, A., Solanki, V. K., & Rani, S. (2022, December). AI-based botnet attack classification and detection in IoT devices. In 2022 *IEEE International Conference on Machine Learning and Applied Network Technologies (ICMLANT)* (pp. 1–5). Salvador: IEEE.

Rahman, M. S., Chamikara, M. A. P., Khalil, I., & Bouras, A. (2022). Blockchain-of-blockchains: An interoperable blockchain platform for ensuring IoT data integrity in smart city. *Journal of Industrial Information Integration*, 30, 100408.

Rajasekaran, A. S., Azees, M., & Al-Turjman, F. (2022). A comprehensive survey on blockchain technology. *Sustainable Energy Technologies and Assessments*, 52, 102039.

Rani, S., Bhambri, P., & Kataria, A. (2023). Integration of IoT, big data, and cloud computing technologies. In *Big Data, Cloud Computing and IoT: Tools and Applications*. USA: CRC Press.

Rani, S., Kataria, A., Chauhan, M., Rattan, P., Kumar, R., & Sivaraman, A. K. (2022). Security and privacy challenges in the deployment of cyber-physical systems in smart city applications: State-of-art work. *Materials Today: Proceedings*, 62, 4671–4676.

Rejeb, A., Keogh, J. G., & Treiblmaier, H. (2019). Leveraging the internet of things and blockchain technology in supply chain management. *Future Internet*, 11(7), 161.

Sahoo, S. S., & Chaurasiya, V. K. (2023). VIBE: Blockchain-based virtual payment in IoT ecosystem: A secure decentralized marketplace. *Multimedia Tools and Applications*, 1–26.

Shah, K., Sheth, C., & Doshi, N. (2022). A survey on IoT-based smart cars, their functionalities and challenges. *Procedia Computer Science*, 210, 295–300.

Sharma, A., Kaur, S., & Singh, M. (2021). A comprehensive review on blockchain and internet of things in healthcare. *Transactions on Emerging Telecommunications Technologies*, 32(10), e4333.

Singh, S., Rathore, S., Alfarraj, O., Tolba, A., & Yoon, B. (2022). A framework for privacy-preservation of IoT healthcare data using federated learning and blockchain technology. *Future Generation Computer Systems*, 129, 380–388.

Surekha, N., Sangeetha, R., Aarthy, C., Kavitha, R., & Anuradha, R. (2022). Leveraging blockchain technology for internet of things powered banking sector. In *Blockchain Based Internet of Things* (pp. 181–207). Singapore: Springer Singapore.

Trnka, M., Abdelfattah, A. S., Shrestha, A., Coffey, M., & Cerny, T. (2022). Systematic review of authentication and authorization advancements for the internet of things. *Sensors*, 22(4), 1361.

Xiao, Z., Li, Z., Yang, Y., Chen, P., Liu, R. W., Jing, W., . . . Goh, R. S. M. (2020). Blockchain and IoT for insurance: a case study and cyberinfrastructure solution on fine-grained transportation insurance. *IEEE Transactions on Computational Social Systems*, 7(6), 1409–1422.

Yin, J., Xiao, Y., Pei, Q., Ju, Y., Liu, L., Xiao, M., & Wu, C. (2022). SmartDID: A novel privacy-preserving identity based on blockchain for IoT. *IEEE Internet of Things Journal*, 10(8), 6718–6732.

13 Decentralized Finance
Toward the Evolution of FinTech

Md. Tauseef and Manjunath R. Kounte

13.1 INTRODUCTION

A few decades back, finance was not seen as a technology-backed industry. The finance industry had its own snail's pace to process the ledger transactions maintained with ink and paper to manual data verifications. The domain established and maintained on continuously evolving economic regulations had numerous loopholes that could be potentially miss-utilized. Most often than not, scams and frauds were the only events leading to significant economic regulation updates, which is still the case in most countries. Even in recent times, where finance has widely accepted technology on its backbone structure, the need for a major paradigm shift toward a new, improved way of transacting is ever-growing. Though forensic accounting is evolving to identify and prevent potential frauds and scams, research shows that the percentage of organizations affected by major scams in a country like India is at 11% only in the context of regulatory and compliance breaches. On the other hand, financial transactions and thereby world economy mostly suffers from over regulations and tight rules by a central authority. Though these rules and regulations are meant to provide safety, they often compromise it with ease of business and thus the region's economic growth. It is no doubt that, in a capitalistic social structure where everything has a monetary value, there should be a mechanism to ensure trust, transparency, and safety. This mechanism, by default, is assumed to be taken by the region's government. However, it is worth mentioning that there have been several instances where the governments have failed to provide stability to their currency. World governments have been and currently are massively investing in advancing FinTech to make the economic system safer and more robust. The advancement in FinTech has contributed to advancements in other sectors such as big data and data analytics, to name but a few (Gai et al., 2018). However, these advancements in FinTech do not intend to solve the most fundamental problem with finance in general—its centralized structure. This centralized structure has three main drawbacks: (1) As the system is dependent on its government to maintain it, it is prone to multipoint failure based on the decisions taken by these governments. (2) The monetary institutions of the world's major governments are used to Quantitative Easing (QE), where the money supply is increased to stimulate a weakening economy by the central banks. The central banks can inflate the money supply whenever they feel necessary. This

DOI: 10.1201/9781003460367-13

trend has at least been observed since the 2008 financial crisis. (3) The structure gives central banks authority to confiscate or exclude money whenever required. There have been several instances where central banks of some countries confiscated money from the bank accounts of its citizen to pay the debt. It is well-understood that the central financial authoritarian body is run by governments and is prone to the basic human behavior of the decision-makers in these bodies. Thus, the requirement for a decentralized financial structure can be well justified.

While finance structure faces the governance issues discussed earlier, it cannot be transformed entirely quickly. Financial technologies aim to solve the procedural problems of the traditional financial industry based on advancements in information technologies. Bank services such as money transfers, debit and credit card services, ATM services, etc., are supplied and supported by FinTech organizations that provide their services to financial institutions to improve their customer service. On the other hand, the emergence of blockchain has highly increased the discussion of its requirement in upgrading traditional ledger technology in FinTech toward a Decentralized Ledger System (DLS). The emergence of blockchain increases the potential of the current FinTech solution industry and introduces a new way of financing system known as Decentralized Finance (DeFi). A more detailed discussion on current FinTech systems, blockchain technology, and DeFi is provided in the upcoming sections (Bali et al., 2023).

13.2 BACKGROUND

The first concept of a decentralized structure in finance was discussed in 1998 (Dai, 1998). It is well-known by now that the limelight on Decentralized Finance (DeFi) fell when Satoshi Nakamoto published the white paper on Bitcoin as a peer-to-peer financial system (Nakamoto, 2008). When published, the system's main aim was to remove the financial institutions serving as trusted third parties for all the transactions between two entities and to remediate the inefficiencies they brought to the system. DeFi has gained a tremendous amount of momentum since then. The attempts to create digital currencies range from B-money (Dai, 1998), RPOW, to bit gold (Szabo, 2005). However, these early approaches required a central authority to manage the transactions. The concept that a cryptographic puzzle as a Proof of Work could be created to validate the transaction without a central authority came later. Though there are numerous cryptocurrencies present these days, their underlying architecture is primarily derived or improved from the architecture of Bitcoin.

Although there was no one agreed-upon date when DeFi was born, a few important events may define the timeline of DeFi, as shown in Figure 13.1. The inception of Bitcoin was a key factor in the origin of DeFi. Bitcoin enabled the creation of Ethereum by Vitalik Buterin in 2015. Ethereum is the default blockchain on which major if not all the DeFi applications are built. However, it must be understood that any robust finance system is formed by a money transfer system and other services such as lending and borrowing, derivatives, trading, and funding. Bitcoin was not designed to handle these services with its limited language called' Script. Ethereum was built to overcome the limitations of Bitcoin's Script. It was built on "Solidity" and used the ERC20 token system.

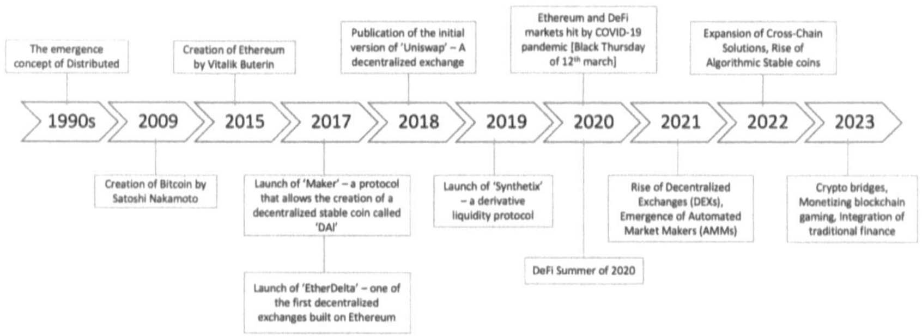

FIGURE 13.1 DeFi timeline consists of all major DeFi growth events until 2023.

Hence, it quickly became the go-to platform for developers worldwide building all sorts of DeFi applications. One of the oldest DeFi projects on Ethereum, "Maker," was launched in 2017 by Rune Christensen. Maker is a protocol that allows the creation of a decentralized stablecoin known as DAI. In the first iteration of the protocol, single collateral DAI only supported ETH as collateral. This was later improved to multi-collateral DAI at the end of 2019. Another project worth mentioning launched around the same time is EtherDelta. EtherDelta was one of the first decentralized exchanges that allowed permissionless ERC20 tokens built on Ethereum. Adding to 2017, one of the important uses cases is ICOs (Initial Coin Offerings). Instead of raising traditional funding to run the project, newly created coins started offering ICOs in ETH for their tokens (Kumar et al., 2022).

Some of today's major platforms that originated in the ICO period are AAVE—a lending and borrowing platform. Synthetix—a liquidity protocol for derivatives, Ren—a protocol for providing access to inter-blockchain liquidity, Kyber network—an on-chain liquidity protocol, and Zero X—an open protocol that enables the peer-to-peer exchange of assets. Another breakthrough was the users interacting with smart contracts containing pooled funds from multiple users rather than interacting directly with a single user. This user-to-contract model was highly suitable for DeFi applications to be built upon. In November 2018, the initial version of Uniswap—an exchange platform—was published to Ethereum mainnet. To the contrast of EtherDelta, Uniswap was built on the concept of liquidity pools and automated market makers. In July 2019, Sythetix launched the first liquidity incentive program. This program later became one of the major catalysts of DeFi summer of 2020. On March 12, 2020, the price of Ethereum rapidly collapsed by more than 30% due to the emergence of the COVID-19 pandemic. This was a major hit to the DeFi market and is popularly recognized as "Black Thursday." The coronavirus pandemic greatly reduced the development in the DeFi market similarly as it affected all the other sectors. However, after a considerable time of downfall, the DeFi sector jumped back with major improvements in a short period in 2020. This jumping back period of DeFi is popularly known as DeFi summer of 2020. In 2021, decentralized exchanges (DEXs) experienced tremendous growth. Trading volumes and liquidity surged on platforms, including Uniswap, Sushiswap, and Pancakeswap. Particularly, Uniswap grew to be one of the biggest DEXs, conducting trades for billions of dollars. AMMs like Uniswap and Balancer became more popular in 2021. These protocols enable decentralized trading by using algorithms to set token prices. Users could trade assets without relying on traditional order books, thanks to AMMs which offered liquidity (Dodmane et al., 2023). Additionally, Polygon's (formerly Matic's) Layer 2 scaling solution has grown in popularity and offers DeFi consumers faster and less expensive transactions. NFTs (Non-Fungible Tokens), while not just used in DeFi, attracted attention for their application in decentralized art and collectible marketplaces. In 2022, a rising number of DeFi projects are being hosted by the platforms Avalanche and Solana Surge, which have drawn notice for their high throughput and inexpensive costs (Fröhlich et al., 2022). In the DeFi ecosystem, initiatives like ThorChain and RenVM sought to make cross-chain asset swaps easier. Algorithmic stablecoins like Terra's UST and FRAX, which provide alternatives to conventional stablecoins, also saw growth in the ecosystem (Ray, 2023). The most famous was Ethereum's Merge, in which the industry leader switched from a Proof-of-Work blockchain to a Proof-of-Stake one. The change resulted in a roughly 99% reduction in the Ethereum network's overall energy consumption. The compromise between decentralization and regulatory systems will be examined in 2023. Crypto bridges, a new innovation in blockchain technology, are one intriguing development. Token and coin exchanges are currently limited to the blockchain on which they were created (Rawhouser et al., 2022). Trading has not been feasible across different blockchain platforms, but cross chain technology will change this. Incorporating conventional finance Traditional finance is being integrated with decentralized finance in novel ways. Real-world assets like corporate loans and mortgages can now be converted into cryptocurrency assets, thanks to blockchain technology (Turi, 2023). One instance is the recent partnership between MJL Capital and Archblock, a division of lending protocol TrueFi. They are collaborating to integrate American community banks into the DeFi network. A fascinating new frontier for non-fungible tokens (NFTs) is in-game purchases. Over one billion people have already won, traded, purchased, and sold in-game goods, claims Finextra. The major events of DeFi summer have been shown in Table 13.1.

13.2.1 Financial Technology (FinTech)

Financial technology aims to bring major changes to the traditional banking system. Apart from the digitalization of the banking system, it also focuses on niche areas such as investments, trading, insurance, automation, risk management, and other banking services. The domain includes everything and anything related to technological improvements in the current financial system. The tremendous implementation of FinTech has led to increased scope of utilizing it in various other technologies. Further, as the scope of FinTech itself has expanded, various technologies have been interbred to meet distinctive requirements. Therefore, a perfect bull's-eye definition of FinTech is hard to provide. However, it can be briefly summarized as technologies used to ease, digitize, and expedite traditional banking processes. This also requires significant changes in business models, products, and services. Though major research in FinTech recently is focused on risk assessment and management of user data, it is primarily out of the scope of discussion in this work.

TABLE 13.1
Major Events of DeFi Summer of 2020

Date	Development
January	*Curve* launches Stablecoin-to-stablecoin decentralized exchange
	AAVE launches decentralized money market
February	Total value locked in DeFi Surpaces $1 billion
	Compound releases COMP tokens
	Decentralized insurance provider *Nexus Mutual* pays out its first claim
March	*Uniswap* launches version 2 of its protocol
	Decentralized exchange *Balancer* launches
May	Decentralized trading platform *Synthetix* partners with *Optimism* to release layer 2 scaling demo
June	*Compound* distributes COMP tokens and hands-off governance to token holders
	Balancer releases Bal tokens
July	*USD coin* stable coin crosses one billion supply
	Yern. Finance launches yield farming protocol
	MakerDAO becomes the first DeFi protocol to hit $1 billion in locked value.
	Synthetix transitions to multiple decentralized autonomous organizations
August	*YAM*, a cryptocurrency with an elastic supply launches
	Curve releases CRV token
	AAVE becomes the second protocol to hit $1 billion in total value locked
	Curve becomes the third protocol to hit $1 billion in total value locked
September	*Uniswap* releases UNI tokens
November	*DAI* Stablecoin supply passes 1 billion
December	AAVE launches version 2 of its protocol

Further, from a business perspective, financial institutions do not primarily involve software or technology development. Instead, the tech-based organizations provide or develop the technological framework and platform required to implement and enhance the banking products. A traditional bank majorly operates in two areas: trust and liquidity. The bank acts as an institution providing trusted repositories where users store their value. It majorly profits by managing the funds it collects and providing fund-based liquidity services to its users (Giuliani, 2020). An average FinTech product is generally an interface optimized on a technology stack that provides certain functionalities to improve existing banking products. For instance, "Unifies Payment Interface" (UPI) is a single-window mobile payment system developed by the Government of India. UPI makes money transfer between bank accounts almost instantaneous. Money transfers between bank accounts existed before UPI was developed. However, UPI solved the specific problem of making it faster. Thus, a basic FinTech stack's generalized architecture can be understood from Figure 13.2.

The architecture consists of modules that add their feature to the system. The improvement in each module is a sector for FinTech start-ups and organizations. Any FinTech solution is built on the core platform of the traditional banking system.

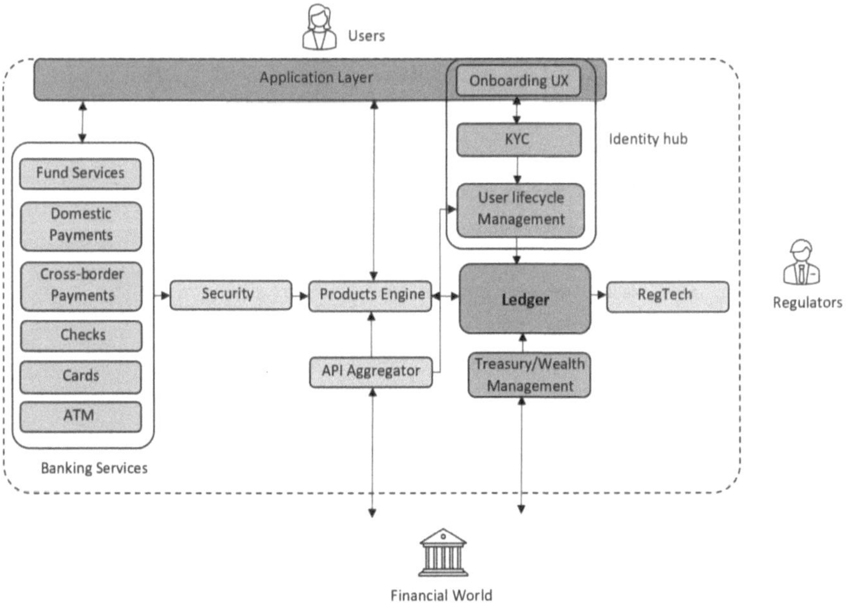

FIGURE 13.2 FinTech service architecture.

The bank or a partnered organization such as Visa or Mastercard offers banking services such as fund services, domestic and cross-border payments, checks, credit and debit cards, and ATMs. These services allow money movement from one account to another account. The identity hub module stores and maintains the user data and provides CRM-based services. The security layer works on top of the banking services layer. It majorly filters out any suspicious and fraudulent transactions. Recent developments in artificial intelligence and machine learning are aiding massively in making the security layer stronger. A product engine is a core product configurator used to configure and disburse banking products such as loan schemes, check services, account services, etc., to customers. This module is the primary functionality banks need to provide their services on a tech-based platform. The ledger module is the first and foremost function of a bank.

All other modules can be considered an upgradation to the functionality of a bank and maintaining a centralized ledger. This module is a database maintained by the bank consisting of all the transactions. A ledger must be highly secure, accessible, reliable, and easy to integrate with other functionalities. The regulation technology (RegTech) module maintains compliance with the country's banking regulations and is updated every time the government passes a new regulation. This module's sole purpose is to make sure that the system is running without violating any regulations passed by the government. API aggregators are used to bring in or provide third-party functionalities to the system. API aggregators are often used to aggregate and provide customer behavior data to other organizations to target their products to the customers. Treasury or wealth management module is used to optimize the management of wealth put by the customers in their accounts to obtain the highest returns from investments while maintaining fund availability to the account holders. Most of the present FinTech organizations either add extra features to this existing architecture or improve the existing modules.

13.2.2 Bitcoin: A General Blockchain Architecture

The architecture of Bitcoin in this section is discussed as presented in Nakamoto (2008). For the sake of discussion, we will refer to a cryptocurrency as a coin. To make the system decentralized, one must first understand a bank's role in the traditional system. The role of the bank can then be replaced. There are four major roles that a bank plays in the current financial system as a trusted third party between two entities that want to transact values (Sudevan et al., 2021).

1. Keeping the users' account information in account balance and ownership.
2. Maintaining a ledger of all the transactions.
3. Preventing double-spending of the same value/coin.
4. Verifying transactions and updating account information.

A DeFi system uses various methods to carry out a bank's works mentioned earlier. The first issue to be addressed is the ledger problem, that is, who will maintain the ledger if there is no central party. Bitcoin solves this issue by making everyone on its blockchain network keep the latest copy of the ledger of all the transactions that ever happened on the network. This way, every entity knows all the transactions on the network and agrees to maintain trust in this shared ledger. Thus, Bitcoin establishes consensus among all its users.

Further, there is one major problem with this approach. If everybody on the network knows all the transactions, the privacy of individual users will be at stake. Bitcoin solves this issue with public and private keys. Public keys can be thought of as cryptographic key that addresses a user's wallet. A private key is used to verify a transaction between two entities. The public keys of users are known to the blockchain network. However, the ownership of a particular public key remains anonymous. Thus, if a third entity wants to track all the transactions of an individual, it can easily track the transactions of a particular public key, but the third entity will not come to know who the owner of the public key is. The only person who knows the owner of a public key is his (/her) self. Further, all the wallets are protected by irrecoverable passwords. In this sense, if the wallet's password is lost by its owner, there is no way it can be recovered, and the Bitcoins in the wallet remain unclaimed forever. Thus, most cryptocurrencies, including Bitcoins, are said to be pseudonymous.

Proof of Work: Traditional banking system controls double-spending by assigning a serial number to each coin. The serial numbers are issued and are controlled by the central bank. The central banks also prohibit simultaneous processing of the same serial number toward multiple receivers, preventing double-spending. Early digital currencies such as B-money relied on the shared ledger only to prevent double-spending, that is, if one person (B) receives a coin from a sender (A), it will be announced to the entire network, and thus the same coin cannot be sent by the same sender (A) to a second person (C). However, this method cannot be applied practically as the time between issuing a transaction (between A and B) and propagating the same to the shared ledger must be considered in a practical scenario. In the above case, the reader must note that the transaction verification happens only between sender (A) and receiver (B) and propagates it to the shared ledger in the network. Bitcoin solves this problem by letting the entire network verify the transaction's legitimacy. As long as a majority of the participants on the network accept the existence and legitimacy of a transaction, the coin will reach the receiver (B).

All new transactions on the network are relayed to each network participant. Once the transactions are verified valid by a considerable majority, they are grouped in a "Block." A timestamp is added to every transaction to make it immutable. Each block contains its index and hash of the previous block and transactions (Kataria et al., 2022). Due to the presence of the previous block's hash, each block in the network can be traced back, and the transactions within each block can be validated. Thus, forming the chain in "Blockchain." The process of adding a block to the chain is performed by all the network participants and is popularly known as "Mining" of the blockchain. It essentially involves calculating a cryptographic puzzle known as a hash. Bitcoin uses SHA256 as its cryptographic hashing algorithm. As every miner in the network is trying to mine a block, there comes a possibility that a particular block can be mined by multiple miners, and the blockchain may branch out eventually, where a single transaction may potentially exist in multiple branches. It becomes impossible to determine which branch to trust. Further, as Bitcoin also awards its miners for successfully mining a block, multiple miners must be rewarded for mining the same block if mined multiple times. This possibility is a major threat to the integrity and stability of the blockchain. To mitigate this threat, there should be a mechanism to decide which block should be added to the chain. This is taken care of by increasing the computational complexity of finding a hash for the block. Thus, the probability of a miner successfully mining a block is directly proportional to the miner's computing capacity. A difficulty level called NONCE (Number Only used Once) is introduced in the block to increase the computational complexity. The nonce dictates that the block should be hashed so that the hash must be lower or equal to a specified value, and the nonce is unknown to the miner. The network accepts only the block hashed as per the nonce requirement. Hence, the miner tries to calculate the hash by incrementing the nonce value every time until a suitable hash is obtained. This process is computationally challenging. Further, the difficulty of the hashing process is determined by the target value set by the nonce. The lower the target value, the fewer solutions exist, and the more difficult the process becomes. For example, if the target value needs the hash (in binary form) to begin with 30 consecutive zeros, it takes 2^{30} attempts to solve the required hash. With the increase in complexity, multiple miners' probability of the same block decreases and is subjected to their computational capacity. Further, Bitcoin includes various other methods to accept a block that multiple miners mine with hypothetically the exact computational capacity. One such way is to consider the longest branch available such that the branch with the highest computational effort is considered. Figure 13.3 shows the structure of each block in the blockchain.

Further, double spending can still occur in certain situations. Since any given transaction is validated by the majority of the participants on the network and then

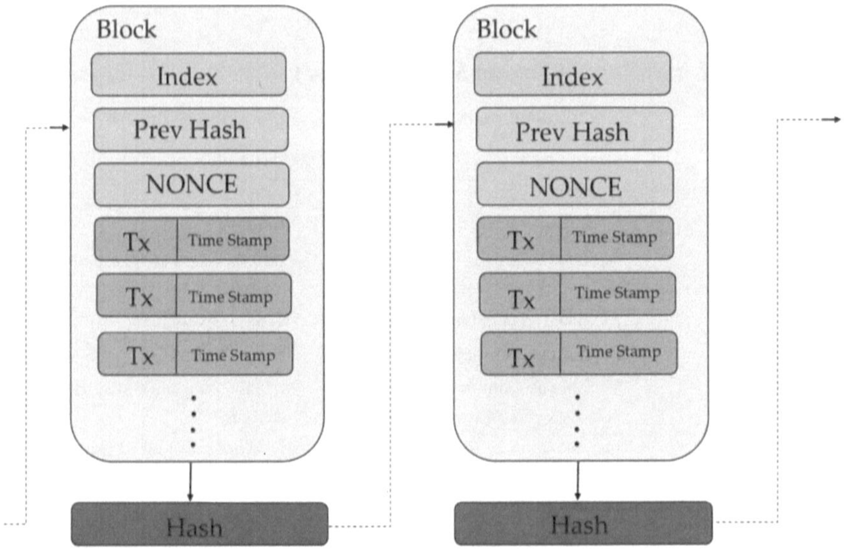

FIGURE 13.3 Structure of a block in the blockchain. The previous block's hash is added to the present block, and thus a chain is maintained.

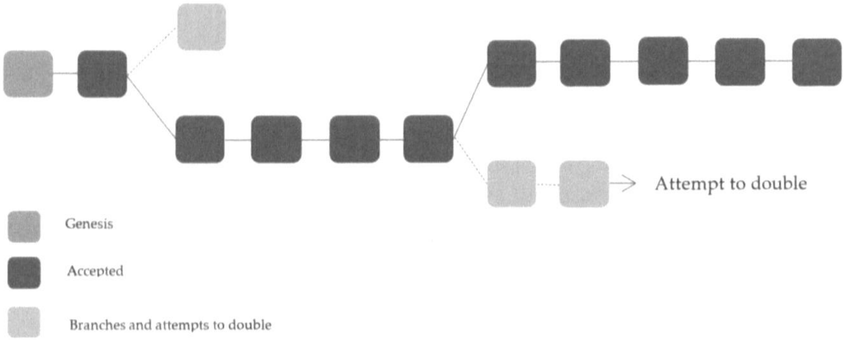

FIGURE 13.4 Branching (forking) of the chain concerning accepted blocks constituting the blockchain.

is added to a block, if a participant gains more computational power than all other participants on the network combined, that is, if a participant gains computing power of more than or equal to 51% of all the computing power of the entire network, she/he will be able to validate a fraudulent transaction and add it to the chain and consequently hash all the blocks in the existing chain according to the changed fraudulent data and calculate Proof of Work to all the blocks. Figure 13.4 shows an attempt to double-spend concerning the original chain of blocks. As this participant controls 51% of computing power, the chain must be accepted by everyone else. This attack on the blockchain is also referred to as a 51% attack. However, any participant gaining

TABLE 13.2
Comparison of Traditional Banking System versus Blockchain System

	Traditional Banking System	Blockchain System
Regulation	Central bank	Consensus
Transaction verification	Centrally	Consensus
Value creation	Debts	Mining
Currency value	Exchange rate	Proof of Work, supply–demand, and trust
Money transfer	Mediated, reversible	Direct, irreversible
Privacy	Implementation-dependent	Pseudonymous
Charges	Variable account maintenance charges and transaction charges	Virtually constant transaction charges
Transaction delay	Instantaneous to few days depending on the mode of transaction	In the order of tens of minutes for all transactions

51% of the global Bitcoin computing capacity is hypothetical. Further, to compensate for increasing hardware capacity in the semiconductor industry, the difficulty level is increased every two weeks.

Blockchain has seen many updates and has grown massively in technology and social acceptance from its early beginning. Undoubtedly, technology has disrupted many industries and has tremendously impacted various other technologies since the internet. It can now be seen as a major enabler for the evolution of Web 3.0. The difference blockchain brings to traditional banking in trust, transparency, and safety must be highly appreciated. The major differences are discussed in Table 13.2. A good technical discussion on blockchain technology has been carried out by many researchers, where most of the technological concepts and challenges of Bitcoin are discussed.

Ethereum is the second-largest cryptocurrency after Bitcoin in terms of market capitalization introduces smart contracts in addition to currency transactions. Smart contracts are a piece of code hosted on the Ethereum blockchain network and can be used to establish contracts between two or more entities. Ethereum is the blockchain network where most DeFi systems have been created recently. Though Ethereum was created based on Bitcoin architecture, it has few improvements. One considerable improvement is that Ethereum will be transitioning from Proof of Work to Proof of Stake to manage the mining process in a few years. However, this transition is not yet implemented and is still under consideration and development when penning this work.

Two major features of blockchain truly define the structure of its acceptance:

1. *Immutability:* Timestamping, hashing, and chaining (1)
2. *Decentralization:* Proof-of-Work algorithm and consensus (2)

Thus, blockchain attains four polymorphic structures depending on the need of a problem:

Public/Permissionless Blockchain: Blockchain is used to its highest potential in a public blockchain form. The core ideology of blockchain is to make a system truly public and permissionless, where an entire community owns the blockchain network, and anybody with sufficient resources can join the network. Both the features of the blockchain (1) and (2) are used to their best capacity in this structure. This system relies solely on its users to maintain the distributed ledger. As the users must be incentivized for adding their resources to the network and as decentralization (2) is computationally complex, public blockchain is an energy-intensive system. Most cryptocurrency networks such as Bitcoin and Ethereum fall in this region (Rani et al., 2022).

Private/Permissioned Blockchain: A private or permissioned blockchain is designed to work in a restricted and closed environment (Rani, Kumar, et al., 2023). A single entity owns this type of blockchain. As the aim of private blockchain is to utilize the security and immutable data storage features of blockchain, it does not require to be decentralized, that is, only (1) is effectively utilized and (2) can be dropped off or a different PoW algorithm can be used to maintain the ownership of the network. This type is usually implemented at a smaller scale within an organization. However, decentralization (3) can still be utilized if needed, but the access remains only within the organization. The owner organization enables security, authorizations, permission levels, and accessibility to the network. Due to their smaller size, permissioned blockchains are faster in processing transactions.

Hybrid Blockchain: Hybrid blockchains are implemented where organizations need some of the data to be available to the public while maintaining the ownership of the network. It allows organizations to build a permission-based hybrid system and a permissionless public system. This allows the organizations to control access to specific data on the network and choose what data can be made available to the public. Transactions and records are not made available to the public but can be verified when needed by allowing access through a smart contract. Hybrid blockchain fully uses (1) and mostly adapts consensus from (2). Thus, the data remains immutable even to the organization by allowing users to verify the correctness of the data. Organizations that need to establish trust with the customers usually set up a hybrid blockchain network (Rani, Bhambri, et al., 2023).

Consortium Blockchain: A consortium blockchain is much like a private blockchain, but more than one organization owns and manages the network. This semi-decentralized system contains multiple organizational nodes which control the consensus procedures. In addition, it contains a validator node that receives, initiates, and validates the transactions. The member nodes can receive and initiate transactions but cannot validate them. This type of blockchain is best suitable for entities such as banks and government organizations.

13.3 DECENTRALIZED FINANCE (DEFI)

The fundamental idea behind the creation of blockchain technology was to remove the intervention of a third-party organization in a transaction between two entities and build a platform for peer-to-peer transaction enablement. The creation of

Ethereum added an extra feature of smart contracts to the blockchain technology, to say the least. However, blockchain at a fundamental level can be considered as an efficient, high-level, decentralized ledger system (DLS) which can be used to store any type of values that require to be immutable in a system that needs to be managed by a network and not by a centralized organization. With this understanding, we can easily infer that blockchain replaces the traditional ledger module in Figure 13.2. However, the outreach of this technology goes far beyond the ledger system toward economical requirements. In the least case, blockchain itself is one of the most secure data storage methods we know to date. Various sectors of industries are starting to accept blockchain as a technology widely. The DeFi ecosystem has massively grown in recent years. DeFi is now offering more services that the traditional market could not offer with added features. It must be clarified here that various types of blockchain are discussed in the previous section. The current DeFi system is purely a public, permissionless blockchain. While there are chances that a decentralized finance system can be created upon other types of blockchain, the extent of their decentralization comes to a question. This argument, however, is left to future developments in technology.

DeFi is built on multilayered architecture. The layers are built on top of each other in hierarchical form. This architecture is highly open and compossible and allows everyone to improve, build on top of, and use some parts of the architecture to create a new application or improve on the existing one. The five layers depend on their below layers to provide functionality and security (Schär, 2021). Also known as DeFi stack, this architecture mainly represents fully public, permissionless blockchain-based DeFi, as shown in Figure 13.5. The individual functionality of each layer can be understood as follows:

Layer 0/Settlement Layer: The settlement layer holds the blockchain network and supports the native asset protocol of the respective blockchain. For example, for Ethereum—a widely used blockchain in DeFi—the native asset protocol would be ETH. The blockchain network in this layer is the fundamental module on which the entire DeFi ecosystem can be built. This majorly acts as a distributed ledger system (DLS) which stores the ownership and transaction data securely and ensures that any change in the data state is followed according to the predefined rulesets of the network. Layer 0 acts as a settlement and dispute resolution layer, that is, if any dispute occurs on the above layers concerning the state of a particular data, the above layers can refer settlement layer to verify and validate the authenticity of a data and its state thus, settling the dispute.

Layer 1/Asset Layer: The asset layer holds all the digital assets issued on top of layer 0. This layer includes the native protocol asset and assets issued on the blockchain network in tokens. Fungible and non-fungible tokens come into the asset layer and are used to track the ownership of a particular asset in the system. One of the significant tokens that Ethereum uses is the ERC20 token. Ethereum Request for Comment (ERC)—20 tokens were implemented in 2015 and are used to implement Fungible tokens (FTs) such as currency, voting tokens, and staking tokens. ERC-721 is the token standard used for non-fungible tokens (NFTs) such as any form of data, including voice, text, image, video, or any other multimedia file. The main difference between FTs and NFTs is that FTs can be interchanged for an agreed-upon value, and NFTs cannot be interchanged (Rani, Kaur, et al., 2023).

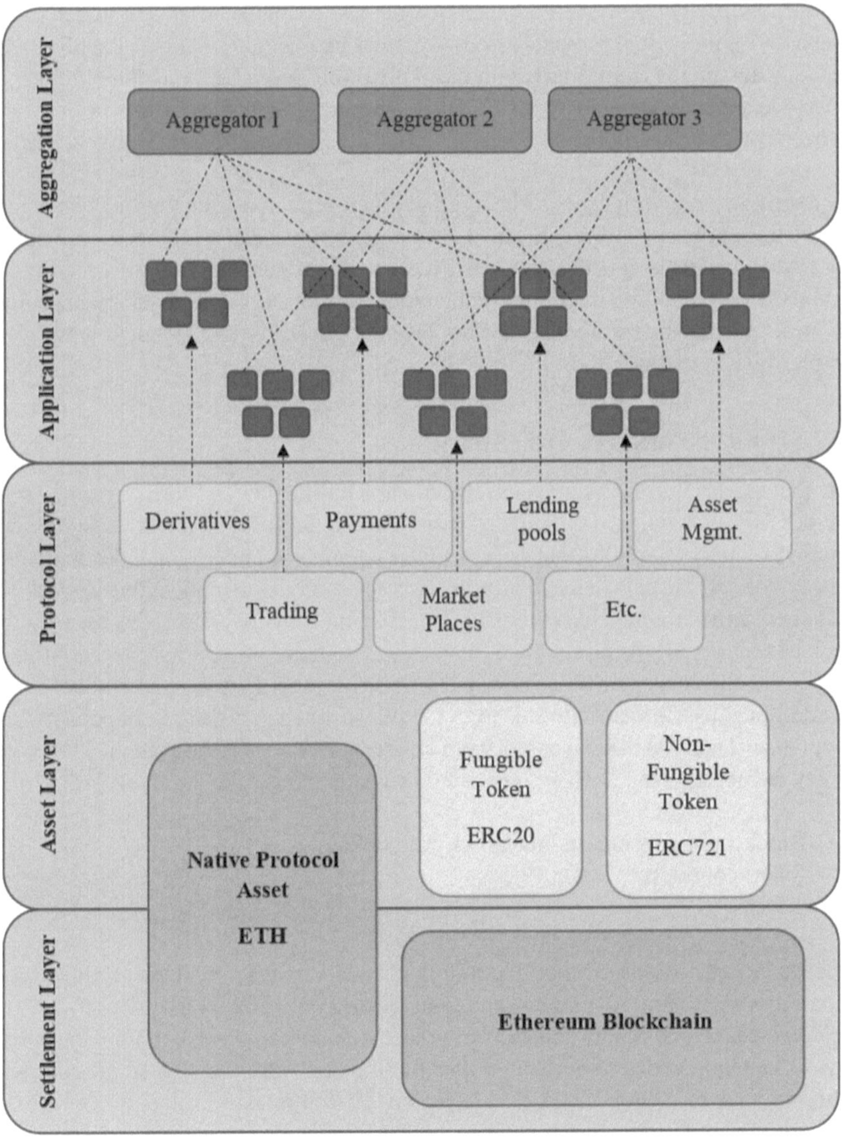

FIGURE 13.5 DeFi system architecture. Each layer depends on the below layer to derive the functionality in the hierarchical layered architecture.

Layer 2/Protocol Layer: The protocol layer consists of all the standard protocols implemented for individual use cases such as decentralized exchanges, marketplaces, asset management platforms, lending pools, derivatives, etc. These protocols are usually implemented on smart contracts and can be accessed by any user or DeFi application. These smart contracts are mostly interoperable and can be built to support any use cases or functionalities, for that matter. Some of the major DeFi protocols are AAVE, yEarn, Synthetix, Compound, etc.,

Layer 3/Application Layer: The application layer provides user-oriented applications built on individual protocols from layer 2. The interaction of these applications is usually abstracted from smart contracts with a web-based UI for a better user experience. Decentralized applications (Dapps) are emerging widely on this layer. Some of the most popular applications are Uniswap, Chainlink, Kyber Network, Radar, Gnosis, etc.

Layer 4/Aggregation Layer: The aggregation layer can be thought of as an extension of the application layer. Layer 4 connects with multiple protocols and applications to provide a specific service. These services can range from providing a dashboard to managing multiple applications, wallet services, asset management services, investment services, etc. This layer is generally employed to simplify a complex task at the user end.

13.3.1 Key Elements of DeFi System

Now that the structure of DeFi is known, understanding how the entire system works together is important. There are three major functionalities that a DeFi system must contain to support the full functioning of the system: The underlying blockchain network to store all the data, a mechanism to write protocols for applications to work on, and a mechanism to track the ownership of assets on the network. As discussed in the previous section, all the protocols in the DeFi system are written in Smart contracts. Further, when blockchain was introduced, it only tracked its native protocol asset, but additional assets were introduced to the network when it grew in popularity and acceptance. These additional assets were created as tokens on the system. Therefore, the key elements of the DeFi system can be considered as follows:

- Blockchain network and its native protocol asset
- Smart contracts
- Tokens

We have already discussed the functioning of blockchain in previous sections. This section discusses smart contracts and tokens (Puri et al., 2022, December).

Smart Contracts: Smart contracts are pieces of code that exist on the blockchain networks and are coded to perform a specific function. The origin of smart contracts went way back to 1994 by Nick Szabo (Szabo, 1996) and was designed to eliminate the requirement of a third party to maintain and execute a contract. These contracts' sole purpose is to activate automatically when certain conditions are met. Therefore, these are simply "if, else, and then statements." As simple as the smart contracts are, they can be used to perform various actions such as releasing funds to appropriate parties of the contract, providing insurance services, registering the ownership of an NFT, and, as discussed earlier, writing protocols on which the DeFi system works (Rani, Pareek, et al., 2023, February). As Ethereum's base language, smart contracts are written in solidity and run on a specific address on the Ethereum network. The basic structure of a smart contract contains data and functions and is anatomically similar to JavaScript and Python. Most often than not, these contracts require triggers to activate. These triggers can be a time, a transaction, another smart contract, or real-world data. Ethereum smart contracts rely on what is known as "Oracles" to provide authentic and verified data to be used as triggers. While there are several challenges still present with smart contracts, the advancements in the field are constantly trying to overcome these challenges (Zheng et al., 2020). These contracts can potentially change the way we look toward law and governance (Raskin, 2016; Jaccard, 2018). While we do not discuss in depth smart contracts, one must appreciate the irony in the fact that we as a global society are moving from "In people, we trust" to "In code, we trust."

Tokens: Apart from non-fungible tokens, the sole purpose is to maintain non-interchangeable data. Fungible tokens are massively used in the DeFi system. These fungible tokens allow for the interchangeability of asset values between various DeFi platforms. Simply speaking, fungible tokens are denominations of a cryptocurrency used to represent a tradable asset and allow its holder to use it for certain purposes (Rani, Kataria, et al., 2023). The economic purpose being the most popular in the general public, there are several types of fungible tokens present on the DeFi system, and they can be classified as follows.

- *Platform Tokens:* Platform tokens use blockchain infrastructures to supply Dapps with unique uses. They benefit from the blockchains they construct upon, acquiring improved protection and the capability to support transactional activity.
- *Security Tokens:* Security tokens represent traditional market securities such as bonds, debentures, options, shares, etc. In general, they are tokens that act as direct blockchain representations of traditional market securities (Bhambri et al., 2023).
- *Transactional Tokens:* Transactional tokens are used to transact in value. They function like devices of account and are exchanged for goods and services (Kataria et al., 2023, July). These tokens frequently function like conventional currencies but, in some cases, offer additional advantages (Singh & Rani, 2023).
- *Utility Tokens:* Utility tokens are added to an existing protocol on the DeFi system and are utilized to access the services provided by the protocol. They often have a symbiotic relationship with the platform they are integrated upon. The platform provides security to the tokens, and the tokens provide the network activity required to strengthen the platform's economy.
- *Governance Tokens:* As the DeFi system grows, the number of decentralized autonomous organizations (DAOs) also increases. Governance tokens are used to provide the decision-making authority of the DAO to its holder. They function much like shares in the traditional market, where the shareholder gets to vote on certain decisions of the organization (Rani, Mishra, et al., 2023).

Now that the functioning of the DeFi system is understood, discussing current use cases of the DeFi market becomes relevant. The decentralized finance market has grown tremendously in the past few years. As a result, the use cases are expected to grow and reach all industries in the upcoming future. Table 13.3 discusses the major use cases present in the DeFi market and compares them with traditional market scenarios (Tanwar et al., 2022, September).

TABLE 13.3
Existing Ecosystem of DeFi Systems and Comparison with Their Traditional Banking System

Use Case	Traditional Finance System	DeFi System
Asset management	Central authority can confiscate or exclude the assets of an individual if needed. As a result, most of the assets are prone to theft	Only the person owning assets can manage/use them as required. Digital assets on blockchain are mostly immune to thefts
Compliance and KYT	The government enforces compliance. Know-your-customer (KYC) guidelines regulate fraudulent monetary activities and terror financing	DeFi enables compliance by participants' addresses rather than that of identity. Also known as know-your-transaction, this helps mitigate the real-time risk and protect against financial crimes
DAOs	Do not exist in traditional finance systems	A Decentralized Autonomous Organization (DAO) is an organization that is run on transparent rules which are encoded on the blockchain network. DAOs eliminate the requirement for centralized administration of the organization
Derivatives	The value of the derivative in the traditional system depends on financial assets. However, these assets are only government-approved assets and do not include cryptocurrencies	Blockchain network-based smart contracts allow the creation and management of tokenized derivatives. The value of these derivatives depends on the performance of the underlying agreements written in smart contracts. These derivatives can represent real-world assets such as bonds, commodities, stablecoins, and other cryptocurrencies
Developer and infrastructure tooling	Development in the traditional finance system is done by a single organization subjected to a central authority's regulations	Composability in DeFi protocols allows various system components to connect, collaborate, and interoperate easily. As a result, the DeFi system allows a strong network rather than a single organization developing the network. The entire community builds and improves on what others have developed
Decentralized exchanges	Currency exchanges may manipulate the price of the currencies. These exchanges charge high exchange rates and are prone to thefts and hackings	Decentralized exchanges operate without the regulation of a central authority where the assets are not in the custody of these exchange platforms. These allow users to exchange cryptocurrencies in a peer-to-peer network and maintain control of their funds. Further, these cryptocurrency exchanges reduce the risk of price manipulation, theft, and hacking

Decentralized identity	A centralized identity requires collaterals such as a users' land, home, income source to check the creditworthiness for providing certain financial services	Blockchain-based identity systems increase the reach of the global economy to a diverse set of users who were not allowed in the traditional system. In addition, DeFi systems reduce requirements toward collaterals for individuals as the users' creditworthiness is determined by reputation and financial activity rather than assets such as home, land, or income source
Insurance	Centralized insurance organizations often cheat their users by declining their claims on illogical terms and conditions, thus increasing their profits	The insurance is coded in smart contracts, and organizations cannot change/manipulate the terms and conditions for their profit. Insurance is automatically disbursed depending on the conditions specified in the smart contracts
Lending and borrowing	Banks control lending and borrowing in the traditional system. For example, banks usually restrict the opportunity to borrow depending on the creditworthiness of an individual	Interest rate markets on peer-to-peer lending networks allow users to earn compound interest on crypto that they supply to the lending pool. The compound smart contract automatically matches lenders and borrowers and calculates interest rates based on the ratio of borrowed to supplied crypto assets
Margin trading	In traditional finance, margin traders leverage their trades by borrowing funds from their brokers. This leverage then forms collateral for the loan	Non-custodial, decentralized lending protocols power DeFi margin trading
Marketplaces	The value exchange in the traditional marketplace contains banks as a third party	DeFi protocols allow a peer-to-peer exchange of products and services globally without the intervention of a third entity
Payments	Banks and central authorities are required for handling and regulating payments	Peer-to-peer payment is the foundational use case of the DeFi system and the blockchain ecosystem in general
Savings	Banks and other financial institutions handle savings	DeFi system contains a peer-to-peer lending pool. This pool can also act as a saving account. It gives a higher return than the traditional bank accounts depending on the dynamic interest rate, which is determined by the supply and demand of the currencies
Stablecoins	Government minted coins are considered stable when compared to the volatility of cryptocurrencies	A stablecoin is any cryptocurrency mapped to a stable asset or group of assets, such as government currencies, gold, or other cryptocurrencies. Stablecoins make blockchains a payment option by reducing the volatility associated with cryptocurrencies
Synthetic assets	Do not exist in traditional finance systems	Like stablecoins, synthetic assets are blockchain-based assets that provide exposure to other assets such as fiat currencies, gold, and cryptocurrencies
Tokenization	Do not exist in traditional finance systems	A token is a digital asset created, issued, and managed on a blockchain. Tokens are designed to be instantly transferable and secure, and they can be programmed with various functionalities depending on requirements (e.g., NFTs)

13.3.2 Major Use Cases of DeFi System

1. *Stablecoins:* Stablecoins are Ethereum tokens that are pegged to a reserve asset such as US dollar. Since they are attached to an asset in the traditional finance system, they are well-known for their decreased volatility (Chauhan & Rani, 2021). Supported stablecoins are subject to the same volatility and risk associated with the underlying asset. Few well-known types of stablecoins based on the type of asset backing are as follows:
    a. *Commodity-Backed:* These stablecoins are backed by real world commodities such as gold. The holders can redeem the coins in exchange to the commodities.
    b. *Fiat-Backed:* The amount of stablecoins of this type is based on the value of the supporting currency, which is held by a financial company controlled by a third party. In this setting, reliance on a supportive asset is essential to the stability of the stablecoin price (Bhambri & Rani, 2024).
    c. *Cryptocurrency-Backed:* Cryptocurrency-backed stablecoins are issued by cryptocurrensets as collateral, which is thought to be similar to fiat-based stablecoins. However, an important difference between the two designs is that although fiat collateralization is usually derived from blockchain, the cryptocurrency or crypto asset used to retrieve this type of stablecoins is made into blockchain, using smart contracts in a more fragmented manner.
2. *Decentralized Exchanges:* Decentralized exchange (better known as DEX) is a peer-to-peer market place where transactions take place directly between crypto traders. DEXs satisfy one among crypto's core opportunities: fostering monetary transactions that are not officiated by means of banks, agents, or any other type of third party. In transactions made through DEXs, the usual third party that generally controls the security and transfer of assets is replaced by a blockchain-based distributed ledger. Some common operating methods include the use of smart contracts or order book relays, but many other options are possible with varying degrees of decentralization.
3. *Synthetic Assets:* Synthetic assets are essentially tokenized derivatives. In the traditional financial world, derivatives are stocks or bonds that traders do not own but want to buy and sell. If you want to profit from the price movements of stocks that you don't basically own, you can do so with derivatives. Synthetic assets, or tokenized derivatives, take this process a step further by adding an entry for the derivative to the blockchain and essentially creating a cryptocurrency token for it. Synthetic assets allow investors to make tokens and trade anything. By using derivatives to bind value in an existing asset and create a token of this exit, investors can easily trade anything on the blockchain. One of the main reasons why synthetic assets become a popular investment method is because of increased security and tracking. While traditional trading takes place in medium trading, with commodity goods, all transactions take place on the blockchain. This assures traders both their anonymity, if they wish to remain anonymous, and their safety, as all transactions are recorded in a distributed ledger.

4. *Money Market DeFi Protocols:* Seeing that DeFi has changed conventional banking structures, borrowing and lending protocols have additionally grow to be one of the main programs of DeFi. several DeFi projects such as Compound and PoolTogether focus at the peer-to-peer (P2P) borrowing and lending marketplace. Disbursed ledger era (DLT) has made transactions faster, greater so within the case of cross-border payments where the fee of transactions and delays prompted bottlenecks for each of the senders and receivers. DLT has democratized banking with the aid of permitting absolutely everyone to take loans or even lend fiat against cryptocurrency collateral. further, the DeFi atmosphere has facilitated tokenization in which digital assets may be created, issued, and controlled on a blockchain network. This has created a brand new form of financial system. For instance, digital assets are being tokenized in the shape of NFTs to create, save, or alternate cost. The boom in DeFi adoption has brought about the boom of DeFi-based totally prediction platforms where users can change cost by way of forecasting the final results of destiny activities.
5. *Insurance Platforms:* One of the most impactful use cases of DeFi has been in the insurance industry. While the present-day insurance machine suffers from complicated audit systems, paperwork and bureaucratic claiming strategies, using clever contracts may want to make it lot greater efficient. DeFi projects along with Nexus Mutual, Opyn and VouchForMe, also provide coverage insurance for cryptocurrency on the blockchain network. The growing costs of inflation and reducing hobby rates in fiat currencies have made savings and investments hard for middle-magnificence people across the world. With no-loss saving techniques, DeFi tasks like PoolTogether, Dharma, and Argent have created opportunity answers for risk-unfastened savings and investments.

13.4 DISCUSSION

A decentralized finance system can be set up in any of the polymorphic forms of blockchain technology. However, these polymorphic forms of the technology, apart from the truly public blockchain (which currently is the only form of DeFi), originate from a basic requirement that organizations need to own and conceal some data from the public to be able to either be profitable or maintain the figure of control and authority in some cases. The current financial system and the governance structure of the world are not designed for all the data and control authority of a large financial or governance system to be completely available to the public. There are chances of a system evolving as a truly public and democratic system. There are equal chances that the system itself will be misused for unethical and unlawful activities. Thus, an overseeing authority/control mechanism must exist to maintain the system's integrity and prevent misuse. Though blockchain technologies have boomed a lot in the last decade, the complete potential these platforms have paved the way for people to use them for illegal activities such as dark web financing and terror financing is still unknown. While world governments are still trying to debate the complete acceptance of blockchain-based decentralized systems, the system is rapidly growing in acceptance and adoption by the world community. However, most of the world's governments are concerned about the legalized adoption of DeFi systems in the main market. The DeFi system undoubtedly removes the government's authority, but it also raises concerns regarding the regulations' ethical utilization of these systems. Finally, it boils down to a fundamental question: "Are governments willing to let go of their authority over their financial system? And if yes, how, what, and where is the mechanism of maintaining the lawful, moral, and ethical integrity of such system to maintain a proper social/economical order?" This truly is an area where new start-ups and organizations can come into the picture. On the other hand, any blockchain and DeFi system is a knowledge or skill-based system, that is, it requires its users to understand the know-how of the platform and requires some extent of technical skills to operate it. As of 2021, only 3.47% of the world population (of 139 major countries) is actively participating in blockchain-based finance technologies. The wide-scale adaption of the DeFi system still requires major evolution in terms of ease of access and simplicity to reach the general public. Further, there will always exist a conflict between world governments which are mostly the opposition to current DeFi systems, and world communities that are trying to prohibit the government's involvement in current DeFi systems. Therefore, for a DeFi system to be successfully accepted and operated worldwide, both parties must come to a common understanding of where a community can operate the system. The government can look over the system's integrity and maintain regulations to some extent. This new form of DeFi can be called Democratic Decentralized Finance (D2Fi). Further, the energy required to run the current blockchain networks is tremendously high. The PoW algorithm is increased in complexity to maintain fair competition between the miners. The increased complexity needs increased computational capacity. This, in turn, increases the energy consumption of the entire system. The worldwide cumulative electricity consumption of Bitcoin alone only in December 2023 is about 146.08 TWh (*CBECI*), and it is constantly growing. The transition of the Ethereum network from Proof-of-Work to Proof-of-Stake algorithm is a promising step here and may reduce energy consumption by some degree. Also, blockchain enablers such as 5G need to be investigated (Dinesh et al., 2019; Tauseef et al., 2023). Further, several criteria define the growth of blockchain-based DeFi systems as a major disruptive technology of the decade. Some of such criteria are discussed in Table 13.4.

13.5 CONCLUSION

Innovation and disruption in financial technologies are seeing more demand than ever before. New ideas result in the evolution of Fintech by bringing a constant state of flux and competing upgrades. As a result, technology has disrupted almost the entire market, and the focus has shifted to user experience. The adoption of AI is a game-changer that provides insights into customer behavior and needs. While most financial institutions try to build trust, massive frauds during online transactions are unwarranted for customers. In addition, cybersecurity breaches and access to sensitive information pose a significant challenge for fintech companies. Managing regulatory risk and compliance adds to the burden.

TABLE 13.4
Criteria Defining the Success of Blockchain-Based DeFi Systems

	Good	Bad	Ugly	Criticality
Privacy	Highly secure since it is distributed ledger with multiple copies across the network	Enterprises with IPs and sensitive information are reluctant	Pseudonymity supports criminal activities, synthetic identity	High
Regulations	Blockchain-based systems can be new governance systems competing with legacy systems	Regulatory regimes cannot keep up with the rapid development in the blockchain and crypto sectors	Lack of regulations leads to scam projects, market manipulations, dark web financing, and cyber ransoms	High
Technology	Employs timestamping to provide immutability of data.	Coding flaws or loopholes, Lack of universal standards. Inefficient technological design	Knowledge of blockchain functionality, applicability, and usage is unclear	High
Trust & Control	Improved data and process integrity, trust, and control of confidential information	DeFi systems rely on the community to establish trust in them.	Cryptocurrencies can be used for dark web financing and terror funding	High
Information Technology	Quality of data provides deep insights to user requirements, multiple access	Finding skilled people for developing a blockchain is a tough task	In the DeFi protocol stack, all the layers are dependent on the layers below them. Thus, if the bottom-most layer is compromised, the entire system is compromised	Medium
Sustainability	One country cannot impose currency restrictions as no third parties or intermediaries exist	The power consensus mechanism (PoW) is energy-hungry	Miners use 0.2% of total global electricity	Medium
Ease	Documentation on a public ledger increased visibility into real-time tracking, transparency to the public, and ease of verification	Public ledger may disrupt organizations'/individuals' privacy	The DeFi systems are still vulnerable to 51% attacks. The immutability myth	Medium

(Continued)

TABLE 13.4 *(Continued)*
Criteria Defining the Success of Blockchain-Based DeFi Systems

	Good	Bad	Ugly	Criticality
Complexity	Employs cryptography and encryption to provide high-level security	Increased complexity demands high computing power and energy resources	Huge complexity in setup and use may demand the replacement of existing systems	Medium
Perception	Blockchain-based DeFi systems will replace traditional, closed (limited services) banking systems	Lacks public acknowledgment and marketing	Lack of central authority to control the havoc if resulted	Low
Speed and Scalability	Seamless transactions across multiple domains	Unable to accommodate large-scale users at a time. When users increase, the network slows down, resulting in huge transaction fees	Lack of cooperation among various stakeholders. Scalability trilemma	Low
Skills	500% increase in demand for blockchain developers, Use of blockchain as a service (BaaS)	Insufficient blockchain literacy	Monopoly of blockchain organizations	Low

With all these challenges, FinTech sector is heading toward a massive disruption of the Decentralized Financial system with blockchain as the enabling technology. With some primary adoptions in the pharma and supply chain, DeFi has gained tremendous momentum and broader awareness in recent times. Blockchain-based DeFi provides ease of documentation, security with cryptography and encryption, privacy with anonymity, transparency with public access, and a democratic currency system. But as a disruptive technology, DeFi has its limitations. While most of the population still perceive cash transactions as a trusted mode and are reluctant to go digital, bringing them to DeFi and migrating the baking systems to a democratic, open banking system is a mammoth task. Lack of awareness, skill gap, computing cost and complexity, scalability issues, and trust among users and financial institutions must be overcome for the intended transformation in the financial sector. In addition, major issues like misuse of Bitcoin as a currency in the dark web and terror funding have to be addressed before we replace the legacy systems with DeFi models. While unbanked and underbanked populations are still significant, massive adoption of DeFi remains a distant dream.

Providing awareness, building skilled resources, and reducing the migration cost with innovation in green technology play a paramount role in its adoption. Furthermore, with ease of access, and seamless services, backed by the solid willingness of the government with perfect regulations and adaptable compliances, DeFi will bring the bank to the customer with varied services, from banking to insurance and from loans to investment. With DeFi, the future seems bright, rich, and wealthy.

REFERENCES

Bali, V., Bali, S., Gaur, D., Rani, S., & Kumar, R. (2023). Commercial-off-the Shelf Vendor Selection: A Multi-Criteria Decision-Making Approach Using Intuitionistic Fuzzy Sets and TOPSIS. *Operational Research in Engineering Sciences: Theory and Applications*, *6*(2).

Bhambri, P., & Rani, S. (2024). Challenges, Opportunities, and the Future of Industrial Engineering with IoT and AI. *Integration of AI-Based Manufacturing and Industrial Engineering Systems with the Internet of Things*, 1–18.

Bhambri, P., Rani, S., Balas, V. E., & Elngar, A. A. (Eds.). (2023). *Integration of AI-Based Manufacturing and Industrial Engineering Systems with the Internet of Things*. USA: CRC Press.

Chauhan, M., & Rani, S. (2021). Covid-19: A Revolution in the Field of Education in India. *Learning How to Learn Using Multimedia*, 23–42.

Dai, W. (1998). *B-Money*. http://www.weidai.com/bmoney.txt.

Dinesh, B., Kavya, B., Sivakumar, D., & Ahmed, M. R. (2019, April). Conforming Test of Blockchain for 5G Enabled IoT. In *2019 3rd International Conference on Trends in Electronics and Informatics (ICOEI)* (pp. 1153–1157). India: IEEE.

Dodmane, R., KR, R., NS, K. R., Kallapu, B., Shetty, S., Aslam, M., & Jilani, S. F. (2023). Blockchain-Based Automated Market Makers for a Decentralized Stock Exchange. *Information*, *14*(5), 280.

Fröhlich, M., Waltenberger, F., Trotter, L., Alt, F., & Schmidt, A. (2022, June). Blockchain and Cryptocurrency in Human Computer Interaction: A Systematic Literature Review and Research Agenda. In *Designing Interactive Systems Conference* (pp. 155–177). South Africa.

Gai, K., Qiu, M., & Sun, X. (2018). A Survey on FinTech. *Journal of Network and Computer Applications*, *103*, 262–273.

Giuliani, G. (2020). *A Neobank Stack*. https://fintechruminations.com/2020/09/08/a-neobank-stack/.

Jaccard, G. (2018). *Smart Contracts and the Role of Law*. University of Geneva, Faculty of Law, Department of Private International Law. Available at SSRN 3099885.

Kataria, A., Agrawal, D., Rani, S., Karar, V., & Chauhan, M. (2022). Prediction of Blood Screening Parameters for Preliminary Analysis Using Neural Networks. In *Predictive Modeling in Biomedical Data Mining and Analysis* (pp. 157–169). New York, NY: Academic Press.

Kataria, A., Puri, V., Pareek, P. K., & Rani, S. (2023, July). Human Activity Classification Using G-XGB. In *2023 International Conference on Data Science and Network Security (ICDSNS)* (pp. 1–5). Tiptur: IEEE.

Kumar, P., Banerjee, K., Singhal, N., Kumar, A., Rani, S., Kumar, R., & Lavinia, C. A. (2022). Verifiable, Secure Mobile Agent Migration in Healthcare Systems Using a Polynomial-Based Threshold Secret Sharing Scheme with a Blowfish Algorithm. *Sensors*, *22*(22), 8620.

Nakamoto, S. (2008). Bitcoin: A Peer-to-Peer Electronic Cash System. *Decentralized Business Review*, 21260.

Puri, V., Kataria, A., Solanki, V. K., & Rani, S. (2022, December). AI-Based Botnet Attack Classification and Detection in IoT Devices. In *2022 IEEE International Conference on Machine Learning and Applied Network Technologies (ICMLANT)* (pp. 1–5). Salvador: IEEE.

Rani, S., Bhambri, P., & Kataria, A. (2023). Integration of IoT, Big Data, and Cloud Computing Technologies. In *Big Data, Cloud Computing and IoT: Tools and Applications*. USA: CRC Press.

Rani, S., Kataria, A., & Chauhan, M. (2022). Cyber Security Techniques, Architectures, and Design. In *Holistic Approach to Quantum Cryptography in Cyber Security* (pp. 41–66). USA: CRC Press.

Rani, S., Kataria, A., Kumar, S., & Tiwari, P. (2023). Federated Learning for Secure IoMT-Applications in Smart Healthcare Systems: A Comprehensive Review. *Knowledge-Based Systems*, 110658.

Rani, S., Kaur, J., & Bhambri, P. (2023). Technology and Gender Violence: Victimization Model, Consequences and Measures. In *Communication Technology and Gender Violence* (pp. 1–19). Cham: Springer International Publishing.

Rani, S., Kumar, S., Kataria, A., & Min, H. (2023). SmartHealth: An Intelligent Framework to Secure IoMT Service Applications Using Machine Learning. *ICT Express*, *10*(2), 420–425.

Rani, S., Mishra, A. K., Kataria, A., Mallik, S., & Qin, H. (2023). Machine Learning-Based Optimal Crop Selection System in Smart Agriculture. *Scientific Reports*, *13*(1), 15997.

Rani, S., Pareek, P. K., Kaur, J., Chauhan, M., & Bhambri, P. (2023, February). Quantum Machine Learning in Healthcare: Developments and Challenges. In *2023 IEEE International Conference on Integrated Circuits and Communication Systems (ICICACS)* (pp. 1–7). Raichur: IEEE.

Raskin, M. (2016). The law and legality of smart contracts. *Georgetown Law Technology Review*, *1*, 305.

Rawhouser, H., Webb, J. W., Rodrigues, J., Waldron, T. L., Kumaraswamy, A., Amankwah-Amoah, J., & Grady, A. (2022). Scaling, Blockchain Technology, and Entrepreneurial Opportunities in Developing Countries. *Journal of Business Venturing Insights*, *18*, e00325.

Ray, P. P. (2023). Web3: A Comprehensive Review on Background, Technologies, Applications, Zero-Trust Architectures, Challenges and Future Directions. *Internet of Things and Cyber-Physical Systems*, *15*, 567–584.

Schär, F. (2021). Decentralized Finance: On Blockchain-and Smart Contract-Based Financial Markets. *FRB of St. Louis Review*, *26*, 7823–7834.

Singh, H., & Rani, S. (2023). Flex Sensor Integrated Smart Strap to Verify Correct Wearing of the Face Mask. *IEEE Sensors Journal*, *24*(2), 2020–2027.

Sudevan, S., Barwani, B., Al Maani, E., Rani, S., & Sivaraman, A. K. (2021). Impact of Blended Learning during Covid-19 in Sultanate of Oman. *Annals of the Romanian Society for Cell Biology*, 14978–14987.

Szabo, N. (1996). Smart Contracts: Building Blocks for Digital Markets. *EXTROPY: The Journal of Transhumanist Thought, (16)*, *18*(2), 28.

Szabo, N. (2005). *Bit Gold*. Recuperado de. https://nakamotoinstitute. org/bit-gold/TVer página.

Tanwar, R., Chhabra, Y., Rattan, P., & Rani, S. (2022, September). Blockchain in IoT Networks for Precision Agriculture. In *International Conference on Innovative Computing and Communications: Proceedings of ICICC 2022* (Vol. 2, pp. 137–147). Singapore: Springer Nature.

Tauseef, M., Kounte, M. R., Nalband, A. H., & Ahmed, M. R. (2023). Exploring the Joint Potential of Blockchain and AI for Securing Internet of Things. *International Journal of Advanced Computer Science and Applications*, *14*(4).

Turi, A. N. (Ed.). (2023). *Financial Technologies and DeFi: A Revisit to the Digital Finance Revolution*. Germany: Springer Nature.

Zheng, Z., Xie, S., Dai, H. N., Chen, W., Chen, X., Weng, J., & Imran, M. (2020). An Overview on Smart Contracts: Challenges, Advances and Platforms. *Future Generation Computer Systems*, *105*, 475–491.

14 Enhancing Access Control in Healthcare Blockchain Using LSTM-Based Breast Cancer Detection and Blowfish Encryption

V. Sunil Kumar, Renukadevi S., Somashekhara Reddy D. and Chandramma R.

14.1 INTRODUCTION

By providing a secure and transparent platform for the exchange and storage of medical records, blockchain technology revolutionizes data management in the healthcare sector. Patient record confidentiality and integrity are ensured by this decentralized system, which forbids unauthorized access and tampering. Smart contracts speed up processes like insurance claims and simplify administrative work by enabling automated, trustless transactions. Additionally, blockchain facilitates seamless data-sharing by facilitating communication between various healthcare systems. Through a distributed ledger, medical professionals can access a patient's whole medical history in real time, facilitating better diagnosis and treatment. Despite challenges like regulatory worries, blockchain's potential for the healthcare industry holds promise for improved patient outcomes, security, and efficiency (Kumar et al., 2021).

Deep learning integration with a blockchain-based healthcare system is a cooperative strategy that fuses blockchain technology's security and transparency with artificial intelligence's (AI's) power. Massive amounts of medical data, such as genetic information, diagnostic images, and patient records, can be analyzed by deep learning algorithms to extract valuable insights and patterns. Blockchain technology is integrated with deep learning models to preserve the confidentiality and integrity of this sensitive data. Consent management processes can be automated by smart contracts on blockchain, ensuring that patient data is used ethically and with explicit consent. Moreover, deep learning can enhance predictive analytics for illness diagnosis and treatment planning, providing tailored and efficient healthcare solutions. Decentralized blockchain technology ensures that AI-derived insights are

 DOI: 10.1201/9781003460367-14

safely shared with authorized stakeholders, promoting collaborative and fact-based decision-making in the healthcare industry.

Encryption is necessary in a blockchain-based healthcare system to safeguard sensitive medical data. It is ensured that patient information is secure and confidential by using state-of-the-art cryptographic techniques such as public-key cryptography. Encrypting each transaction and data entry and dispersing the keys across the decentralized network lowers the risk of unauthorized access. This encrypted layer provides additional protection for electronic health records, prescription data, and other sensitive healthcare data. As a result, blockchain encryption enhances confidentiality, integrity, and trust, making it crucial for maintaining the security standards needed by healthcare systems (Liang et al., 2020).

14.1.1 Main Contributions

- Building a secure blockchain for healthcare with Hyperledger that enhances patient data security, secure key management, and efficient BC detection.
- BC detection is accomplished via LSTM. It provides a method for analyzing and identifying patterns with an improved memory.
- A reliable cryptographic method for safeguarding private medical information is blowfish encryption, which provides robust security for information about breast cancer. Because of its strong encryption technology, data security is enhanced by maintaining the confidentiality and integrity of breast cancer–related information.
- The study yields notable results, demonstrating an effective encryption speed of 0.2535 ms and 98.5% accuracy in BC detection when compared to benchmarks, indicating improved security and performance in the proposed blockchain-based IoT system (Miriam et al., 2023).

14.1.2 Organization of Works

Section 14.2 provides a summary of the pertinent works; Section 14.3 provides a brief explanation of the suggested model; Section 14.4 displays the findings and validation analysis; and Section 14.5 provides an overview and conclusion.

14.2 RELATED WORKS

Assiri proposed a modified blockchain concept in an attempt to prevent delays in life-threatening medical conditions. The research looked at healthcare operations and data to classify transactions based on their nature and sensitivity. Rather than using more traditional protocols like Proof of Stake or Proof of Work, a fair Proof-of-Stake consensus protocol was created for block generation and correctness. The project created three different types of blocks, sped up block verification, and categorized transactions into three categories based on timeliness and reliability. This alignment with authoritative tiers and significant healthcare scenarios aimed to reduce computation costs, enhance security, and expedite the trustworthy block and transaction verification process. An overall execution time reduction of 49% was demonstrated when comparing experimental results to conventional blockchain models.

The approach of Al Hwaitat et al. provided a novel method for implementing a light-weight authentication system and optimizing data storage in a massive Internet of Things system based on permissions and blockchain. The method addressed scalability issues and storage issues in blockchain-based systems. Interestingly, the user included homomorphic encryption for the first time to encrypt IoT data before uploading it to the cloud.

In a paper published by Miriam et al., the Lionised Golden Eagle based Homomorphic Elapid Security (LGE-HES) technique for blockchain protection for health networks was initially introduced. Utilizing MATLAB software, the research was conducted while maintaining the protection of medical imaging data by using hash functions. Tests utilizing datasets of magnetic resonance imaging (MRI) and computed tomography (CT) images showed the important implications of the suggested framework, which was 94.9% successful in identifying and categorizing fraudulent communications during simulation.

In their study, Kalapaaking et al. suggested a blockchain-based federated learning (FL) system that uses SMPC model verification to safeguard healthcare systems from poisoning attacks. For encrypted inference processes, FL participants submitted their machine learning models; models that were judged compromised were removed. Verified local models were securely aggregated on the blockchain node, and a variety of medical datasets were used to test the proposed architecture.

According to Selvarajan and Mouratidis, a secure method for exchanging healthcare data is called Consultative Transaction Key Generation and Management, or CTKGM. In this system, distinct key pairs were created using timestamps and multiplicative operations, and they were securely stored on a blockchain. The Quantum Trust Reconciliation Agreement Model (QTRAM) ensured reliable data transport by using trust scores from feedback data. Utilizing trust value and feedback analysis, the platform employed Tuna Swarm Optimization (TSO) to validate nonce verification messages during transmission, facilitating secure patient-system communication.

A convolutional Bi-LSTM network was introduced in the paper by Kolhar and Aldossary for the assessment of time-series medical images that have been fully homomorphic-encrypted (HE). Analytical sequence layers based on bi-LSTM were used to encode time data, and convolutional blocks learned to recognize particular spatial features from hidden image sequences. By employing a sequence voting layer with weighted units and geographical weights, efficiency was raised and the quantity of incorrect diagnoses was reduced. To show the system's efficacy, the public datasets CheXpert and BreaKHis were utilized as strict benchmarks.

14.2.1 Research Gaps

Even though these results significantly enhance the use of blockchain technology in healthcare, research gaps still exist. Usually little research is done on the scalability and applicability of proposed models. Moreover, it remains uncertain how suitable

FIGURE 14.1 Block diagram.

these frameworks will ultimately be to transform healthcare environments. To ascertain how the proposed solutions will be impacted by external factors such as changes in laws or technological advancements, more research is required. Furthermore, extensive validation using multiple datasets and rigorous testing under various conditions are needed to prove the robustness and generalizability of these state-of-the-art techniques. Filling in these gaps would increase the practicality and reliability of blockchain applications in the real healthcare environment.

14.3 PROPOSED METHODOLOGY

Figure 14.1 displays the suggested work flow using the methods employed in this investigation.

14.3.1 Blockchain Technology in the Medical Field Making Use of Hyperledger

In the domains of software-defined networking (SDN), Internet of Things, and smart cities, the blockchain is a crucial instrument. Consider the benefits of blockchain technology in healthcare to ensure security and privacy. The importance of using blockchain technology is to ensure the privacy preservation of the users in an Internet of Things-enabled healthcare system in order to safeguard data. Blockchain integration with healthcare systems addresses ownership, access control, privacy, and security of health data. Smart contracts have benefits over conventional blockchain implementation in terms of privacy preservation. Blockchain is used in the suggested system to increase security and efficiency. The three primary components of an IoT-enabled industrial system are data centers (DCs), full nodes (FNs), and local nodes (LNs). The data can be sent by the LNs to FNs in the edge-blockchain layer because to their limited resources. In addition to being utilized for block addition and mining to the blockchain network, FNs can help LNs search through transactions. Lastly, DCs are in charge of keeping FN data for an extended period of time when needed. In accordance with needs, the DCs transmit information to the periphery. All three nodes are registered using the suggested blockchain-based privacy and security system, which then uses the proposed smart contract–based PoW to authenticate data transactions in the network. Complete transactions are also stored on the Inter Planetary File System (IPFS) storage system, and the blockchain stores the hash string that is generated. Finally, breast cancers are identified using the DL-based privacy and security scheme. Applications for blockchain include fault tolerance, redundancy, and data distribution. Figure 14.2 shows the data's organization and flow.

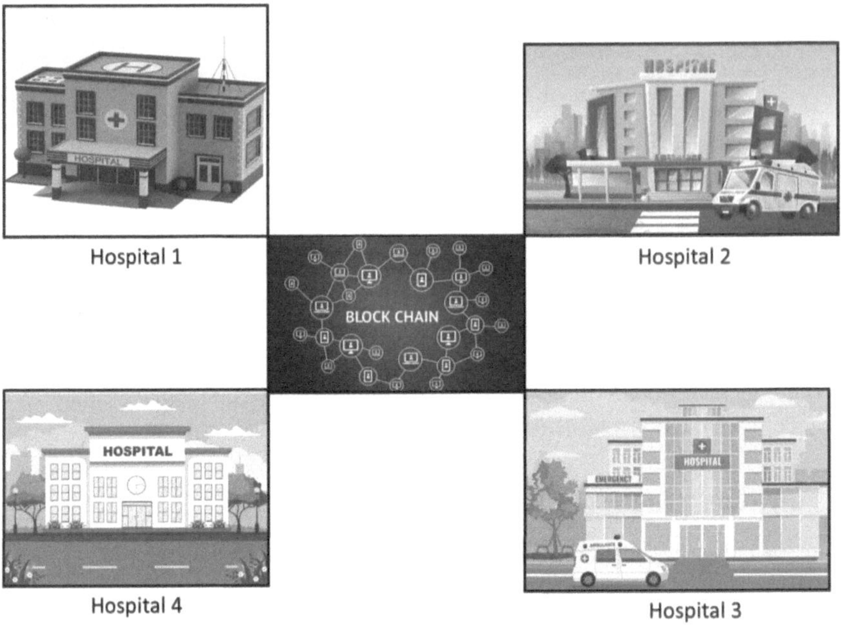

FIGURE 14.2 Blockchain-based cross-domain healthcare system.

14.3.2 Blockchain Technology and Proof of Work (PoW)

One significant example of a decentralized technology that facilitates peer-to-peer communication is blockchain. Blockchain technology removes reliance on centralized nodes.

The consensus mechanism, which is made up of 51% of all nodes, authorizes transactions. The technology known as blockchain is transparent and immutable, meaning any data or transaction recorded on it cannot be altered. Figure 14.3 clearly illustrates how a blockchain network functions and how it is applied in a number of areas, including the IoT, smart grid systems, smart cities, and healthcare systems. Blockchain is a peer-to-peer (P2P) transaction ledger that operates in a decentralized manner and can be employed to safeguard data transmission in the Internet of Things between edge nodes and clouds. Timestamp, relevant data, and the block are composed of the previous block's hash made for a particular transaction. Proof of Work (PoW) is the procedure that guarantees an added block to the blockchain cannot be later removed or changed. The two popular consensus algorithms used to verify a transaction's validity and add fresh blockchain blocks are Proof of Stake (PoS) and Proof of Work (PoW). Nevertheless, both strategies may be breached throughout the complete Internet of Things if a malicious miner gains a more powerful computer or stake than 51%; this is referred to as the 51% attack. Figure 14.4 illustrates in detail how new peers and nodes are added using blockchain technology.

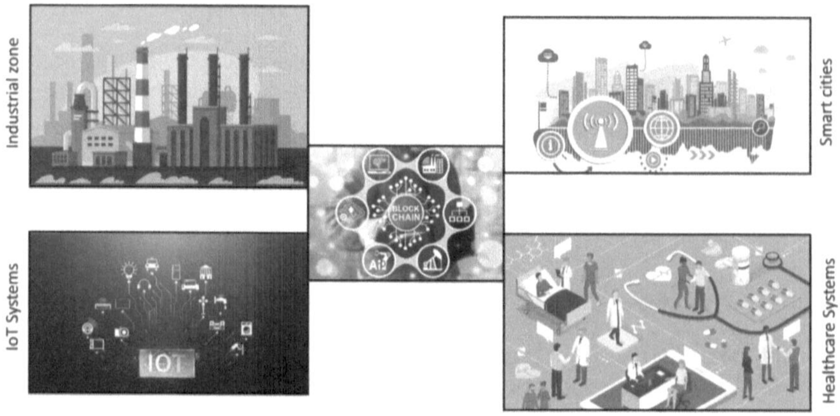

FIGURE 14.3 Blockchain applications in a range of industries.

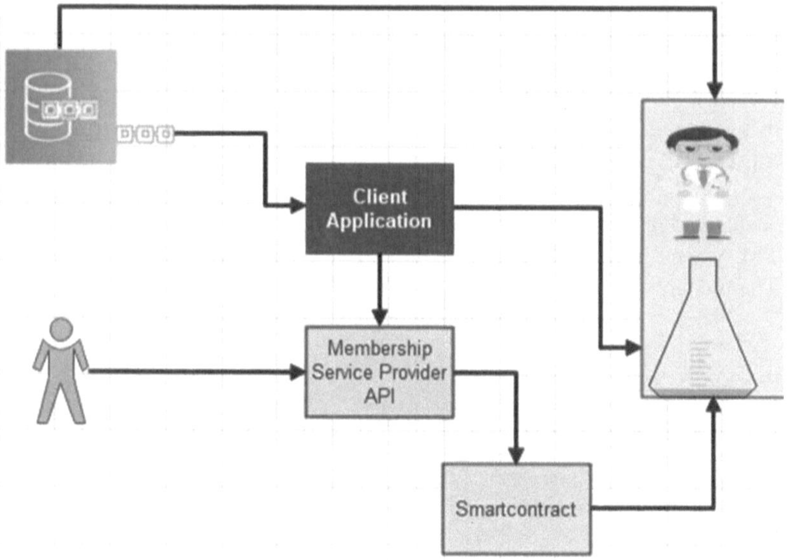

FIGURE 14.4 Electronic health record transactions via a blockchain-based healthcare system.

The functioning of the proposed healthcare systems utilizing blockchain technology and intelligent smart contracts is depicted in Figure 14.4. Deep learning–based smart contracts can recognize and monitor security risks presented by outside sources.

14.3.3 IoT Model with a Proposed Blockchain and DL

The Internet of Things (IoT) is a vast network of interconnected sensors, network devices, and other electronic nodes. Blockchain is crucial to IoT network security because it keeps the hub node disconnected from the network. Figure 14.4 illustrates how the Internet of Things is being used in various industries, including healthcare, smart cities, and sensor devices. The provided dataset is processed using an LSTM model to detect breast cancer, and the results are encrypted for security.

14.3.4 LSTM Model for BC Detection

This chapter uses LSTM for network-wide breast cancer detection. Sequential information data has been generated by RNNs for numerous deep learning applications, including object tracking, image classification, voice recognition, and translation. LSTM and gated recurrent units are two varieties of RNNs (GRUs).

The following are the formulas for recursive RNNs:

$$h_t = \tanh\left(W_h h_{t-1} + W_x x_t\right) \tag{14.1}$$

$$y_t = W_y h_t \tag{14.2}$$

Here x_t is the initial vector, h_t is the hidden layer, y_t is the vector output of the experiment, and W_h is a matrix with weights. To establish a computational environment, gather input, and produce output, the RNN is used with the LSTM. Long-term memory is formed from short-term memory during this process. Three gates make up the LSTM system: input, forget, and output.

The hidden state is computed by LSTM in this way:

$$i_t = \sigma\left(W_f\left[h_{t-1}, x_t\right] + b_i\right) \tag{14.3}$$

$$f_t = \sigma\left(W_f\left[h_{t-1}, x_t\right] + b_f\right) \tag{14.4}$$

$$\text{Sigmoid} = \frac{1}{1 + e^{-1}} \tag{14.5}$$

$$\tilde{o}_t = \sigma\left(W_o\left[h_{t-1}, x_t\right] + b_o\right) \tag{14.6}$$

$$\tilde{c}_t = \tanh\left(W_c\left[h_{t-1}, x_t\right] + b_c\right) \tag{14.7}$$

$$c_t = f_t * c_{t-1} + i_t * \tilde{c}_t \tag{14.8}$$

$$h_t = o_t * \tanh\left(c_t\right) \tag{14.9}$$

Here σ is the logistics' sigmoid function; i, f, and o are the gates for input, forget, and output, respectively; h is a hidden vector with constant size that shows up in all layers; W is a matrix of weights that is employed to transform data from gate vectors to cells; and each gate has a vector-only feature called m that gets its input from the feature m of the cell vector. In Equation (14.7), $\tilde{c}_t$ is a concealed component responsible for the input layer at this time; c_t is the internal memory that this device

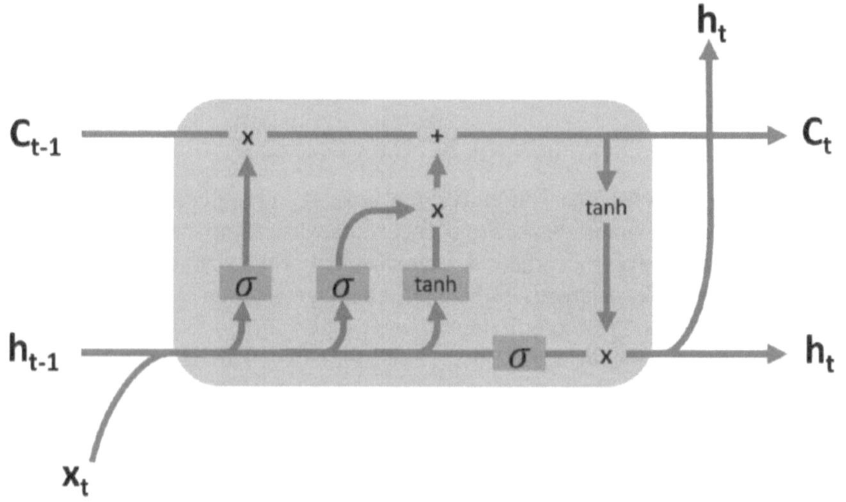

FIGURE 14.5 Architecture of long short-term memory (LSTM).

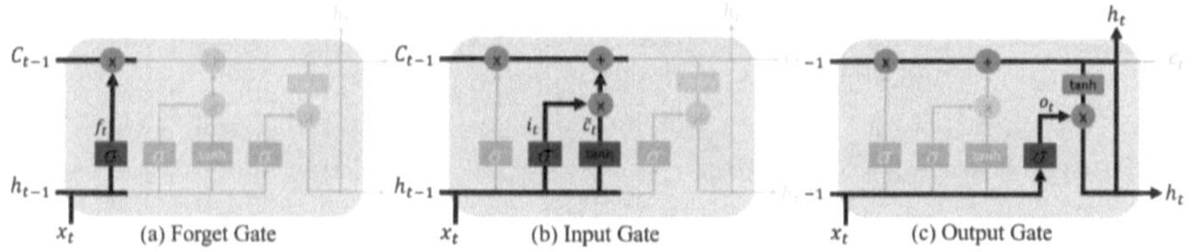

FIGURE 14.6 LSTM gates.

computes; and h_t is the result of multiplying memories to obtain a hidden state, as shown in Figure 14.5.

The forgotten entrance in Figure 14.6a is in charge of eliminating data from the state of the cell. Two inputs are required: the hidden state's output from the preceding time step (h_{t-1}) and the current time step's input (x_t). We add a bias and multiply these inputs by weight matrices. Next, an output vector comprising of values between 0 and 1 is obtained by applying a sigmoid function. This vector is used to determine which values to retain and which to discard.

Subsequently, as illustrated in Figure 14.6b, the input gate uses a two-step process to send information to the cell condition. A sigmoid function is a filter that is comparable to an input gate that is used for h_{t-1} and x_t to construct a values-based vector between −1 and 1 that are appropriate for the cell state. Next, values from the cell state may be enhanced by this vector.

Which data from the cell state are output depends on the output gate, as seen in Figure 14.6c. Three steps are used in the LSTM output gate function. To convert the values between −1 and 1, the hyperbolic tangent function tanh is initially utilized in the state of cells after the vector has been constructed. Next, a filter for values of the previously hidden state is created through sigmoid function application to it: h_{t-1} and x_t. LSTM output information is then produced by multiplying the values filtered by the first step's vector creation.

14.3.5 Blowfish Encryption for BC Data Privacy

This study uses blowfish encryption to safely store the breast cancer data. Sixteen-round Feistel arrangements are used to encrypt data when using the Blowfish encryption algorithm. In every cycle, there is a replacement of the key and the subordinate of the information. These are all 32-bit XOR and augmentation tasks. Blowfish uses a variety of strategies. Prior to utilizing any encryption or decryption methods, these keys need to register in advance. P1, P2, . . ., P18 are the 18 32-bit subkeys that comprise the key clusters, commonly referred to as the P-exhibit.

Four 32-bit S-boxes totaling 256 entries each are present: S1, 0, S1, 255; S2, 0, S2, 255; S3, 0, S3, 255; S4, 0, S4, 1, . . ., S4, 255.

Algorithm 1's function that loops over the network 16 times is necessary for the encryption. Every iteration comprises a permutation that is dependent on data and a substitution that is also dependent on data and key. All operations on 32-bit words are additions and XORs. Every cycle only requires four additional steps: retrieving data banks for the indexing array. The communication instrumental variable x, which has a 64-bit value, is 7 of 14. Gap x is divided into xL and xR, two 32-bit components. The following are the steps that make up the encryption process:

Algorithm 1. Blowfish *F* Function

Divide x into two 32-bit halves: xL, xR
For i = 1 to 16:
xL = XL*X*XRPi
$xR = F(XL)XOR \times R$
Swap *XL* and *xR*
Swap *XL* and *xR* (Undo the last swap)
xR = *xR* XOR P17
xL = *xL* XOR P18
Recombine *xL* and *xR*
For Function F: partition xL into four eight-piece quarters: a, b, c, and
dF (xL) = ((SI, a + S2, bmod232)XOR*S*3, c) + S4, dmod232.

Except for the use of PI, P2, and P18 in the switch configuration, decryption is the same as encryption. The circle should be unrolled and all subkeys should be kept in the cache-store during Blowfish executions that demand the fastest speeds.

14.4 RESULTS AND DISCUSSIONS

14.4.1 Experimental Setup

The trials were carried out on a desktop workstation with 64 GB of RAM and an Intel(R) Core i9–7900X CPU running at 3.30 GHz. The computer was running

Ubuntu 16.04, a Linux-based operating system. We used scikit-learn and PyTorch, two well-known machine learning frameworks, to execute the simulations.

14.4.2 Dataset Description

The UCI ML repository hosts the WDBC dataset, which was created by Dr. William Wolberg of the University of Wisconsin. In this study, it was utilized as a dataset to develop an IoT-based DL model for breast cancer diagnosis. The dataset consists of 569 participants with 30 real-value features and 32 characteristics. On the target's output label, samples classified as benign or malignant are indicated by two classes. A 569 × 32 feature matrix representing the dataset contains 212 malignant and 357 benign patients.

14.4.3 Performance Metrics

The following performance metrics were employed in this chapter:

- *ACC:* "Proportion of the precisely predicted observation to the total number of observations."
- $$T^{\text{accuracy}} = \frac{\left(Tr^{p} + Tr^{n}\right)}{Tr^{p} + Tr^{n} + Fa^{p} + Fa^{n}} \tag{26}$$
- *SEN:* "The quantity of genuine positives that are precisely identified."
- $$Se = \frac{Tr^{p}}{Tr^{p} + Fa^{n}} \tag{27}$$
- *SPEC:* "How many actual negatives there are, which are precisely calculated."
- $$Sp = \frac{Tr^{n}}{Fa^{n}} \tag{28}$$
- *PR:* "The proportion of exactly predicted positive observations to all positively predicted positive observations."
- $$Pr = \frac{Tr^{p}}{Tr^{p} + Tr^{p}} \tag{29}$$
- *F1-score:* It is described as the "the recall and PR harmonic mean. It is employed as a statistical metric for performance evaluation."
- $$F1\text{-score} = \frac{Se \cdot Pr}{Pr + Se} \tag{30}$$

Table 14.1 weighs the BC detection model's performance analysis against the suggested model, and Table 14.2 displays the timings of the different models' execution.

In Table 14.1 and Figure 14.7, various approaches are compared based on performance metrics such as F1-score, specificity, accuracy, precision, and recall. The AE

TABLE 14.1
BC Analysis of Various Models

Methods	Accuracy	Precision	Recall	F1-Score	Specificity
AE	93.8	93.7	93.7	92.8	92.7
VAE	94.8	94.7	94.5	94.4	94.2
ANN	95.9	95.5	95.2	95.3	95.1
RNN	96.7	96.6	96.5	96.3	96.2
Proposed LSTM model	**98.5**	**98.3**	**97.8**	**97.8**	**98.6**

TABLE 14.2
Execution Time Analysis

Models	ENCR Time (ms)	DECR Time (ms)
AES	1.2462	1.1265
RSA	0.7652	0.7431
DES	0.6578	0.6341
Twofish	0.5621	0.4334
Proposed blowfish encryption	0.2535	0.2123

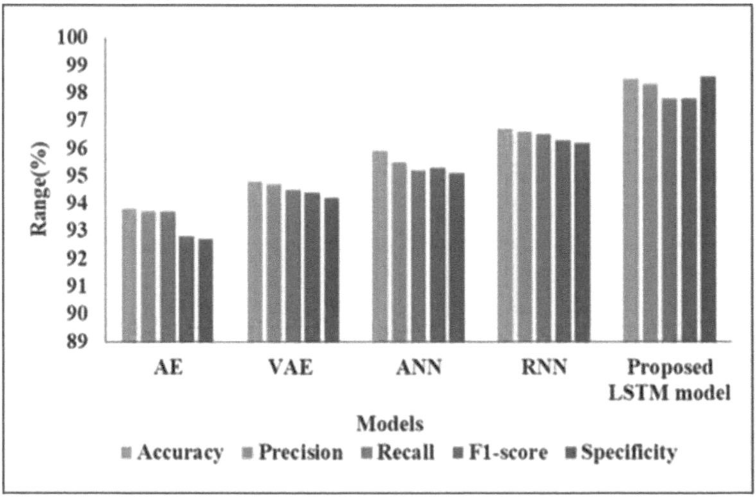

FIGURE 14.7 BC detection analysis of proposed model with existing models.

method produces 93.8% accuracy, precision, and recall with an F1-score of 92.8% and a specificity of 92.7%. With a 94.4% F1-score, 94.8% accuracy, 94.7% precision and recall, and 94.2% specificity, VAE performs better than the average. The ANN method shows even more improvement with accuracy of 95.9%, precision of 95.5%,

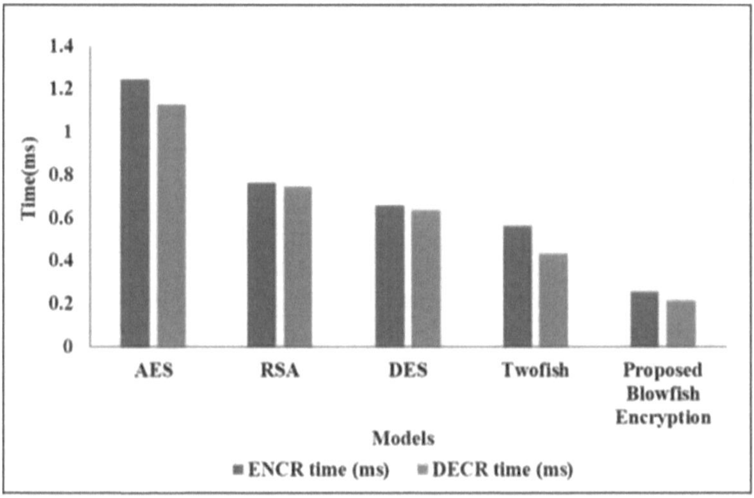

FIGURE 14.8 Execution analysis.

recall of 95.2%, F1-score of 95.3%, and specificity of 95.1%. RNN performs better than previous methods with 96.7% accuracy, 96.6% recall and precision, 96.3% F1-score, and 96.2% specificity. The Proposed LSTM model ultimately outperforms all other methods, displaying 98.5% accuracy, 98.3% precision, 97.8% recall, 97.8% F1-score, and an impressive 98.6% specificity.

Table 14.2 displays, in milliseconds (ms), the encryption and decryption time values for a range of encryption models. AES is one of them, with encryption and decryption times of 1.2462 ms and 1.1265 ms, respectively. For the RSA model, the encryption and decryption times are 0.7652 ms and 0.7431 ms, respectively. For the DES model, the encryption and decryption times are 0.6578 ms and 0.6341 ms, respectively. Twofish requires 0.5621 ms for encryption and 0.4334 ms for decryption. Lastly, in comparison to the other encryption algorithms mentioned, the suggested Blowfish encryption model has the lowest encryption and decryption times, at 0.2535 and 0.2123 ms, respectively. This indicates how efficient the model is. Comparing the proposed model, the decryption and encryption times of the current models are shown in Figure 14.8.

14.5 CONCLUSION

Conclusively, this research addresses the crucial issue of patient health record (PHR) security through the development of an innovative DL-based searchable blockchain system. The study uses blowfish encryption, the Hyperledger tool, and smart contracts to strengthen security for medical data by taking advantage of keyword search vulnerabilities. The proposed method offers unique security measures for secure key updates and revocation, which sets it apart from existing blockchain-based healthcare systems that place a higher priority on data storage. The study highlights the effectiveness of the recommended security framework by demonstrating a strong

98.5% accuracy in BC detection. Long short-term memory (LSTM) is utilized. A dataset from the Internet of Things is used to further validate the suggested access control mechanisms, demonstrating their superiority over benchmark models. Smart contract implementation ensures the practical usability of the proposed methods in a secure, distributed healthcare setting. The results demonstrate the significant advancements in security, anonymity, and user behavior tracking, defining the proposed approach as a game-changing innovation in blockchain-based Internet of Things implementations. The technology optimizes performance and strengthens data security with an astounding encryption speed of 0.2535 ms. The comprehensive solution this study offers for the issues with traditional blockchain-based healthcare systems contributes to the expanding body of knowledge on healthcare security. As technology advances, the proposed study sets a good benchmark for the creation of effective and safe healthcare data management systems in the future. To achieve ongoing development, future work should look into AI-enhanced security measures that are flexible, include quantum-resistant algorithms, and expand the applications to other healthcare settings.

REFERENCES

Kumar, R., Khan, A. A., Kumar, J., Golilarz, N. A., Zhang, S., Ting, Y., . . . Wang, W. (2021). Blockchain-federated-learning and deep learning models for covid-19 detection using CT imaging. *IEEE Sensors Journal*, *21*(14), 16301–16314.

Liang, W., Zhang, D., Lei, X., Tang, M., Li, K. C., & Zomaya, A. Y. (2020). Circuit copyright blockchain: Blockchain-based homomorphic encryption for IP circuit protection. *IEEE Transactions on Emerging Topics in Computing*, *9*(3), 1410–1420.

Miriam, H., Doreen, D., Dahiya, D., & Rene Robin, C. R. (2023). Secured cyber security algorithm for healthcare system using blockchain technology. *Intelligent Automation & Soft Computing*, *35*(2).

15 Transforming Emergency Care

A Blockchain-Powered Smart Admission Monitoring System

Reda Chefira, Radia Belkeziz, Mohamed Yassine Samiri, and Said Rakrak

15.1 INTRODUCTION

In the wake of federal incentive programs that enabled hospitals and ambulatory clinics to install electronic health records systems, there has been a rush to understand what new healthcare technology can accomplish with patient data. Emerging technologies in healthcare like AI and blockchain have the potential to scrutinize patient data exchanges. Blockchain enables cooperating parties with conflicting interests to maintain a distributed, tamper-proof digital ledger. It is valuable for any two-party contract that must be auditable. One of the challenges with medical records today is that they are not always available for medical professionals to access and blockchain is able to fix this problem. One of the most powerful use cases for blockchain lies in smart contracts. Blockchain technology's security and safety make smart contracts and associated digital agreements appropriate for the healthcare industry as they provide an auditable record of all transactions and interactions, increasing transparency. Smart contracts simplify healthcare processes as the management of patient records, the processing of insurance claims, and the distribution of healthcare-related payments. Additionally, they assist in managing the allowance of limited available resources during emergencies, wherein the need for up-to-date and reliable information is critical. Altogether, leveraging smart contracts holds the potential for efficient and transparent healthcare-related processes, potentially enhancing patient outcomes in emergency situations. Moreover, healthcare practitioners may benefit from using AI in several areas of patient care and administrative procedures. There are various AI applications used in healthcare industry:

- One of the most prevalent types of AI in the medical field is machine learning. Large medical datasets are used to extract insights to improve clinical

DOI: 10.1201/9781003460367-15

decision-making and patient outcomes, automate daily tasks for healthcare personnel, speed up medical research, and increase operational effectiveness.

- Natural language processing is the capacity of computers to comprehend the most recent terminology used in human speech and writing. Unstructured clinical notes can be analyzed by NLP systems, providing a wealth of knowledge that can be used to enhance procedures and provide better outcomes for patients.
- The most basic type of AI, rule-based expert systems, use established knowledge-based rules to resolve issues. They are based on several "if–then" principles, which are widely used in the healthcare industry.
- AI has several administrative uses in the healthcare industry. It can be deployed in a wide range of applications, such as managing medical records, revenue cycle management, clinical documentation, and claims processing.

Additionally, smart emergency care represents the cutting-edge of healthcare technology, where the convergence of AI and blockchain is revolutionizing the way we respond to critical medical situations. AI-powered systems now play a pivotal role in assessing patient conditions, optimizing ambulance routes, and even recommending the most suitable hospitals in real time. This ensures that patients receive prompt and efficient care during emergencies. Simultaneously, blockchain technology is securing the entire process by storing patient data and treatment histories in an immutable ledger. This not only guarantees data integrity but also enables rapid and secure sharing of critical patient information among healthcare providers. Together, AI and blockchain are forging a new era of emergency care that not only is efficient but also prioritizes patient privacy and safety. As these technologies continue to evolve, the potential to save lives and enhance emergency medical services becomes increasingly promising.

Hence, by embedding a hospital recommender system for urgent medical situations within a blockchain-based framework, the following major challenges are addressed:

- *Data Privacy:* Patient-sensitive information running on blockchain technology enables it to be securely managed and only available to authorized parties: thereby protecting patient privacy, a key concern in the field of healthcare, leveraging the intrinsic cryptographic strength of blockchain.
- *Real-Time Decision Support:* Potentially improving the speed and accuracy of decision-making in acute medical situations, blockchain and AI merge to provide real-time, data-driven recommendations for hospital selection. The decision-making criteria involved derive from an in-depth analysis of literature approaches along with prospects for further research.

Meeting these challenges, such a framework lays the foundations for a safer, interconnected, and responsive healthcare ecosystem.

In this chapter, we propose a framework based on the convergence of AI and blockchain to ensure the comprehensive care of a patient from their transfer in an ambulance to their admission at the hospital. Upon arrival at the hospital, medical professionals take over to provide the necessary care and update the patient's medical records. This framework employs, on one hand, a recommender system to direct the ambulance to the most suitable hospital, considering the patient's condition and decision criteria, thereby streamlining the decision-making process and ensuring the patient receives appropriate treatment while minimizing the risk of logistical or medical mismatches. On the other hand, the relationship between the patient and their caregivers relies on smart contracts deployed in the blockchain to ensure the security of their data and respect their privacy.

Following this introduction, Section 15.2 moves on to discuss the recent related work. The proposed framework is explained in Section 15.3, in which the design and the methodology are detailed. Section 15.4 reports on the findings regarding blockchain and AI approaches, while also discussing the significance of conclusions based on the results obtained. Finally, Section 15.5 summarizes the overall assessment, outlines open research areas, and sets out potential avenues for further enhancement within the framework.

15.2 RELATED WORK

The integration of AI and blockchain results in possibly the most trustworthy technology-enabled decision-making system ever created, one that is essentially tamper-proof and offers dependable insights and conclusions. In the case of our research, healthcare data-sharing is being improved by merging blockchain and AI. A system created to manage healthcare data is referred to as a "health information system" (HIS). Systems that gather, store, manage, and communicate hospital operational management, systems supporting healthcare policy decisions, and systems handling a patient's electronic medical record all are included in this HIS. Electronic medical record (EMR) systems is defined as "an electronic record of health-related information on an individual that can be created, gathered, managed, and consulted by authorized clinicians and staff within one health care organization" [1]. These technologies can streamline processes, enhance the standard of patient care, and increase patient safety while also providing essential, sensitive information. In [2], authors discussed the necessity of storing information in a decentralized way for providing data and identity security. They also presented the role of AI in electronic health record systems. They proposed the integration of blockchain and AI to manage these systems as AI could be utilized to identify certain efficient prediction models and blockchain may be used to save in a decentralized network critical data and expertise of medical experts along with the credential of employees. An AI-assisted blockchain-based framework is presented in [3]. It is supposed to save doctors' time by enabling them to go through patients' medical records. Authors explain the system's operation through several steps. First, patients upload their prescriptions and laboratory reports. Then, the uploaded media is stored into the Ethereum platform in the form of IPFS hash values. Next step consists of feeding machine learning with data. The data is then prepared, refined, and processed to generate a single summary for each patient using Computer Vision API among other techniques. This summary can be accessed by the doctors and provides them with the patient's medical history. Tagde et al. [4] debated on AI and blockchain integration, where blockchain is allowing access to patients' medical records and AI provides decision-making capabilities. They assume the integration comes with service efficiency, cost reduction, and democratization of healthcare. They discussed immutability, decentralized consensus, baked-in incentives, and security blockchain within the healthcare industry. A blockchain AI framework for healthcare records management is presented in [5]. The framework's requirements are elicited by a goal-oriented modeling approach with the Constrained Goal Model. Five high-level intermediate goals are defined as Enforcing blockchain, Providing Tools, Increasing Security, Reducing Transaction Costs, and Increasing Interoperability and captured various non-functional requirements: New Jobs, Empowering Youth, Improved Public Health, Securing Immediate Family HRs, and Reduced Transaction Time. These goals were set to satisfy the requirements of an electronic healthcare records system where the technologies of AI and blockchain are combined. This article [6] presents a conceptual showcase of a technology framework centered around patients. This framework incorporates AI, blockchain, and wearable technology, and delves into the potential utilization of these combined advancements in the realm of chronic disease supervision. Authors affirm that incorporating these technologies have the potential to enhance current approaches to managing chronic illnesses. In [7], authors seek to explore the domain of blockchain technology, examining its structure, existing uses, amalgamation with novel advancements, and the potential it holds across various realms within healthcare and clinical research. The studies included in the analysis showcased diverse framework designs and a range of applications in the healthcare sector. These applications encompassed areas like diagnosing and managing chronic diseases, as well as monitoring and evaluating health conditions. Their findings suggest that blockchains hold great potential to revolutionize clinical trial management. This potential is evident in applications such as employing smart contracts, allowing participants to control access to their data, establishing trustless protocols, and ensuring data accuracy. This article [8] offers an overview of recent investigations into the fusion of AI technologies and cutting-edge blockchain solutions. These efforts aim to enhance and create novel technical benchmarks within the healthcare system, specifically concerning the exchange of medical diagnostics and electronic health records. Authors state that integrating technologies such as the IoT, big data, cloud computing, and AI leads to the enhancement and modernization of conventional medical systems in terms of efficiency, service quality, and customization, while blockchain technology offers improved data management, access control, and data integrity within the healthcare systems.

In fact, by thoughtfully integrating the predictive capabilities of AI alongside blockchain's promise of data security and transparency, a new era of precision and efficiency in healthcare responsiveness has begun. Inspiring confidence in this trend of scientific research serving to transform emergency healthcare, this merger marks not only a technological advance but also a significant step toward a more responsive and safer healthcare environment for all.

15.3 PROPOSED FRAMEWORK

15.3.1 Overview of the Proposed Framework

The proposed framework features a security-conscious, intelligent architecture in which several parties interact, from the time the patient is cared for by the ambulance unit to the time of admission to the hospital. Figure 15.1 illustrates the framework

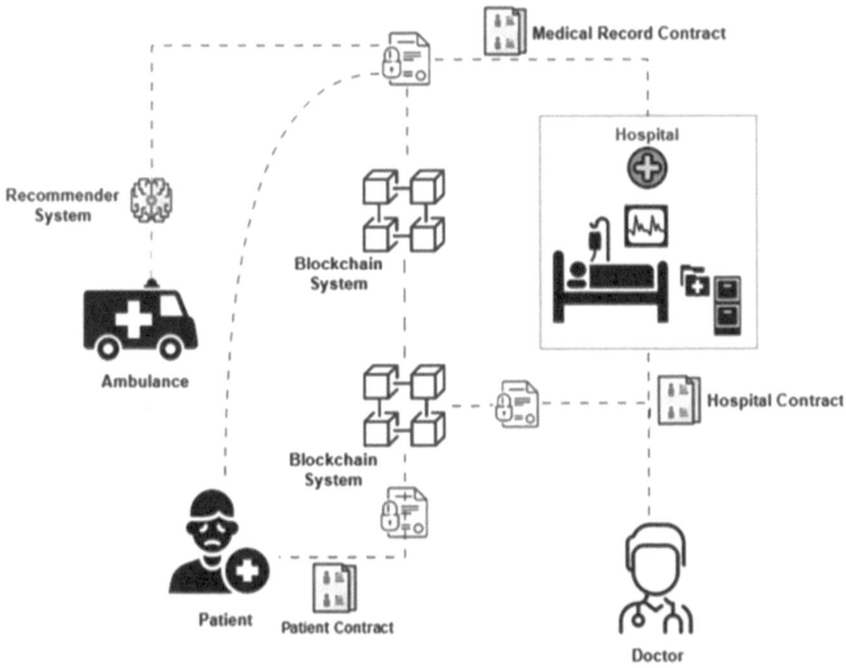

FIGURE 15.1 Illustration of the proposed framework.

while highlighting the actors as well as the processes. It includes the following actors: patient, doctor, ambulance, and hospital. The smart contracts made between these actors are hosted in an integrity- and security-assured blockchain system. In addition, a recommender system enables a hospital to be suggested according to several decision parameters.

15.3.2 Framework Modeling

Figure 15.2 illustrates a well-orchestrated process for delivering efficient healthcare services by seamlessly integrating technology. It begins with the patient in need of medical attention, where the ambulance swiftly takes charge, ensuring prompt initial care. The involvement of a recommender system is a forward-thinking addition, as it not only recommends suitable hospitals but also streamlines the decision-making process. The ambulance's communication with hospitals to verify availability demonstrates real-time coordination and resource optimization.

Figure 15.3 provides a visual representation of the interaction between these two entities. At the onset, the doctor initiates the sequence by sending a data inquiry request to the blockchain, seeking access to a patient's medical records. The request includes essential details such as the patient's identification and the specific data needed. The blockchain system processes this request, verifying the doctor's identity and permissions. Once authorized, the blockchain retrieves the requested data from its decentralized ledger and shares it with the doctor.

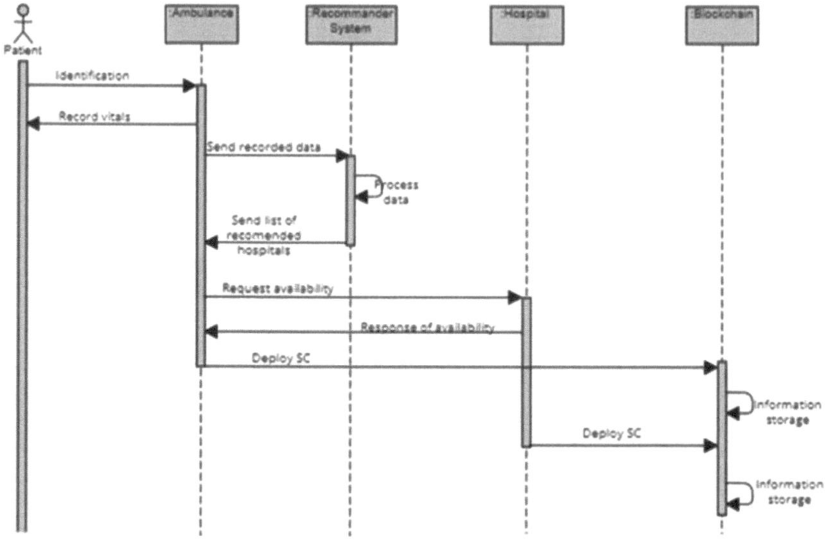

FIGURE 15.2 Sequence diagram of the initial care process.

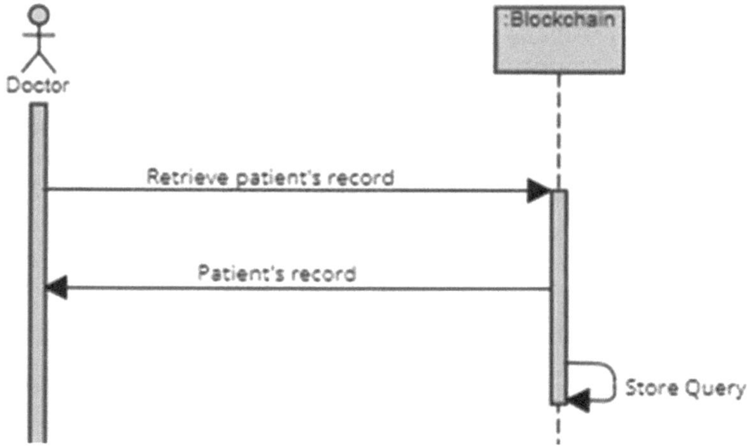

FIGURE 15.3 Sequence diagram of data inquiry and storage between the doctor and the blockchain.

The process commences with the patient's interaction with a doctor or the medical staff, emphasizing the importance of human expertise in healthcare. As treatment is administered, the diagram showcases the seamless integration of blockchain, wherein treatment data and patient information are securely recorded. This blockchain integration ensures that the treatment history remains tamper-proof and accessible to authorized parties. This approach not only fosters trust and transparency in

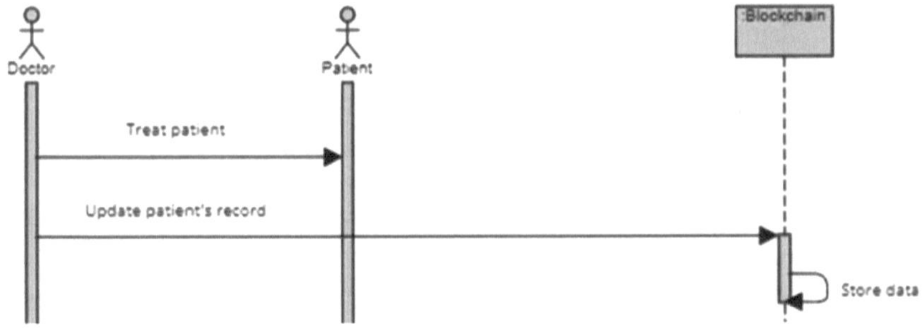

FIGURE 15.4 Sequence diagram of the medical care process.

healthcare but also facilitates seamless collaboration among healthcare providers. Figure 15.4 highlights this process.

15.3.3 AI-Based Hospitals Recommender System

Probabilistic data structures with low clutter serve as an excellent pre-processing tool before conducting a search query in a database. These filters, thus, can be used to authenticate several actors in an authentication-based system. Bloom filter is a space-efficient probabilistic data structure. This approach can be employed for checking if a given entry belongs to a set. For instance, a query about whether a login is in fact part of a collection, where the collection is the list of all registered login names, is a flawless application. And as the method is probabilistic, it is necessary to point out that it can yield some "false positive" results. However, it also never generates "false negative" results; in other words, no login that is not in the set will be accepted [9, 10]. Whenever an application calls for a substantial number of elements to be stored along with striving for as low a false positive rate as possible, cuckoo filters allows less space cost than Bloom's filters [11]. Murmur, FNV series of hash functions, and Jenkins hashes can be used as non-cryptographic hash function for the above filters [12–14].

In this chapter, a cuckoo filter-based AI was chosen as a recommender system to select the most suitable hospital for the patient's condition. The criteria were chosen with a view to building on the innovative features discussed in the state-of-the-art chapter. Below, we review each parameter and discuss how important this is to the recommender system:

(1) *Hospital:* It refers to the identifier of the hospital in each geographical area by means of a set of key parameters, helping to improve efficiency of the model on which the recommendation system process is based.
(2) *Proximity:* It denotes the metric distance between the hospital and the patient's location involved in a medical emergency. Its value is rated on a scale of 0 to 1. Recommending the nearest hospitals potentially suitable for the patient's care contributes to shortening the paramedic's travel time and thus providing efficient access to medical care.

(3) *Emergency Services:* It denotes whether the hospital offers an emergency service, which is essential for treating critical medical cases.
(4) *Triage Capability:* It involves prioritizing patients according to the severity of their condition. With triage capability, hospitals can manage large numbers of patients efficiently in an emergency, allowing them to allocate available resources optimally.
(5) *Specialization for Emergencies:* Specialized hospitals have the appropriate expertise, equipment, and resources to deliver focused and effective care in specific emergency scenarios, such as trauma centers or cardiac care units [15–17].
(6) *Response Time:* It concerns the time the hospital takes to respond to an emergency call or to a patient's admission. Rapid response times are crucial in emergencies, as they lead to faster medical intervention and better outcomes for patients.
(7) *Medical Staff:* It indicates whether the hospital is staffed with suitable medical personnel. Well-trained medical staff in sufficient numbers will be needed to handle several cases at once and provide timely, effective care.
(8) *Facilities:* It relates to the level of availability of medical equipment and resources. Adequately equipped hospitals cope with a wider range of pathologies and offer comprehensive care, making them ideally suited to emergency situations.
(9) *Patient Rating:* It reflects users' assessment of the hospital. Patient ratings provide insight into quality of care and patient satisfaction, thus assisting patients and their families to better anticipate and make informed decisions in emergency situations.

The datatypes of the recommender system parameters are shown in Table 15.1.

All these factors combine perfectly to identify the most appropriate hospital for a patient in an emergency. Proximity, emergency services, specialization, response time, medical staff, facilities, and patient assessments all are essential factors to be considered to ensure that patients receive the best possible care in critical scenarios. Consequently, the recommender system relies on the above list of parameters to enable the learning model to recommend the most appropriate hospital.

TABLE 15.1
Datatypes of the Recommender System's Parameters

Parameter	Datatype
Hospital	String
Proximity; response time; patient rating	Number
Emergency services; triage capability; specialization for emergencies; facilities; medical staff	Binary (true, false)

15.4 RESULTS AND DISCUSSION

15.4.1 Recommender System

15.4.1.1 Score Metrics

The learning model scores are precision, recall, and F1-score. These choices are driven by the following considerations:

- In healthcare models, precision is an important criterion, as it measures the accuracy of positive predictions. High precision implies a high probability of being correct when the model predicts a hospital, therefore avoiding irrelevant recommendations.
- Recall measures of the model's ability to determine all relevant hospitals. It reduces the risk of false negatives, which can lead the paramedic to direct the patient to a hospital that is not suitable in terms of the emergency.
- Achieving a high F1-score is critical, as it indicates a good balance between precision and recall. This means that the model can make accurate positive predictions while retaining a high rate of actual instances.

15.4.1.2 Simulation Results

The simulation begins by defining a dataset in which all the criteria of the recommendation system are initialized. The criteria are then tested to initialize a maximum distance, emergency specialization, emergency services, and medical staff. Once the patient's medical condition has been identified by the paramedics (e.g., trauma), the dataset is analyzed to compute the hospital scores based on the following scheme:

- Hospitals with a distance to the patient of less than max_distance (proximity).
- Hospitals equipped with SpecializationEmergency (true)
- Hospitals offering EmergencyServices (true)
- Hospitals benefiting from Medicalstaff (true)

The better the hospitals meet these parameters, the higher their overall score. The recommender model then compiles the scores of all hospitals and recommends the most appropriate ones for the patient's state of emergency. The system then draws up a representative map of the chosen hospitals (refer to Figure 15.5). In this scenario, Hospital X and Hospital P have the highest scores among all the seven hospitals.

15.4.2 Blockchain System Simulation

For the implementation of the smart contracts: CuckooFilter.sol, PatientContract.sol, HospitalContract.sol, and MedicalRecordContract.sol on the Ethereum blockchain, we used Remix platform. These contracts serve as the backbone for our healthcare solution, leveraging the transparency and security of blockchain technology to streamline patient data management. We also presented detailed tables that provide valuable insights into the deployment and execution of these contracts. Tables 15.2–15.6 highlight information such as gas usage, transaction costs, and

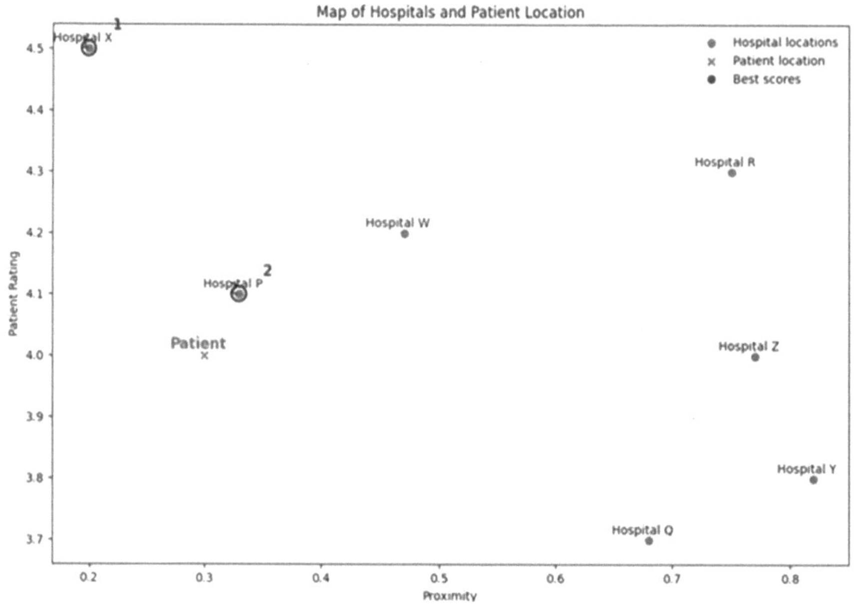

FIGURE 15.5 Representative map of the simulated scenario showing the recommended ones based on the patient emergency situation.

TABLE 15.2
General Information about the Simulation

Account Parameters	Details
Blockchain	Ethereum
Gaz limit	3,000,000
Ether	100

TABLE 15.3
Smart Contract Deployment Details

Contract	Gas	Transaction Cost	Execution Cost
CuckooFilter	558879	486111	401461
Patient	2088699	1816590	1680800
MedicalRecord	2440372	2122484	1964030
Hospital	1750000	1521985	1407919

TABLE 15.4
Patient Contract Functions' Gas Costs

Contract	Gas	Transaction Cost	Execution Cost
addPatient	336930	292982	267990
updateHospitalization	116747	101519	79595
getPatient	–	–	46094(*)
Total	–	394501	–

(*) Applies when called by a contract.

TABLE 15.5
Hospital Contract Functions' Gas Costs

Contract	Gas	Transaction Cost	Execution Cost
addHospital	246309	224849	200793
addMedStaff	141761	123270	100990
getMedStaff	–	–	20699*
Total	–	348 119	–

* Applies when called by a contract.

TABLE 15.6
Medical Record Contract Functions' Gas Costs

Contract	Gas	Transaction Cost	Execution Cost
addPatient	421815	366795	341183
addRecord	389940	339078	314338
updateHospitalization	61688	53641	28017
getRecord	–	–	88220*
Total	–	759514	–

* Applies when called by a contract.

execution costs for every function within the contracts, ensuring transparency and efficiency in managing our blockchain-based healthcare ecosystem. These contracts and their respective metrics enable us to establish a robust and reliable healthcare data infrastructure on the Ethereum blockchain. The general information related to the simulation is depicted in Table 15.2.

15.4.3 Discussion

Based on the recommender system results, we notice that Hospital P is closer than Hospital X; however, the recommender model gives a higher score to Hospital X as the latter's other decision parameters are most suited to the patient's emergency condition. Table 15.7 shows the entry parameters for each chosen hospital by the AI.

Furthermore, Figure 15.6 illustrates the recommender system scores underlying the selection of the two hospitals.

Based on both the proposed architecture and simulated results, it is clear that blockchain technology holds great potential for improving emergency medical care:

- *Greater Record-Keeping:* Blockchain can assist in developing a safer and truer record of patient information, such as medical history, diagnoses, as well as facilitating care plans. Doing so can help decrease overall errors while improving care quality.
- *Strengthened Data Security:* Advanced encryption techniques are used in blockchain technology to secure data, making it difficult for unauthorized individuals to access or alter patient records. In turn, that contributes to maintaining patient privacy and forestalling medical impersonation.
- *Coordination and Communication Efficiency:* During an emergency scenario, quick and easy access to, and the exchange of, relevant patient

TABLE 15.7
Input Parameters Related to AI-Based Recommender System

Parameters	Hospital-X	Hospital-P
Rank	1	2
Proximity	0.52	0.33
Emergency services	True	True
Specialization emergency	True	True
Response time	15	18
Medical staff	True	True
Facilities	True	True
Patient rating	4.5	4.1

```
+------+------------+------------------+
| Rank | Hospital   |      Score       |
+------+------------+------------------+
|  1   | Hospital X |       5.8        |
|  2   | Hospital P | 4.97333333333333 |
+------+------------+------------------+
```

FIGURE 15.6 The scores related to the chosen hospitals.

information among healthcare professionals is crucial. Blockchain aids in easing this process as it provides a decentralized, trusted sharing platform for exchanging data.

- *Cutting Costs:* Getting rid of the need to capture and store data in a hands-on manner, blockchain lowers costs involved in providing healthcare. This in turn may serve toward rendering emergency medical care far more accessible and cost-effective, as shown in Tables 15.4–15.6.
- *Better Veracity:* Blockchain can promote greater transparency by providing an auditable and accessible trail of all transactions and interactions in the healthcare system. Reaching this point helps foster trust and confidence from patients to service providers and improve overall liability in all aspects of the healthcare system.

15.5 CONCLUSION

The focus of this chapter is on exploring blockchain's suitability within the medical industry, particularly in managing patient care during emergency scenarios in a connected monitoring environment with IoT at its core. Leveraging blockchain's secure and transparent framework, the goal is to streamline the process from the moment a patient requires immediate assistance to their admission into the hospital. To achieve efficient and secure patient care, a simulation scenario is outlined, involving patient contracts, hospital contracts, and healthcare record contracts. Through their interactions, the delivery of patient care becomes more streamlined and reliable. In this scenario, whenever a patient requires emergency care, an Emergency Medical Technician (EMT) captures and inputs the patient's data into a blockchain-based system, establishing the groundwork for subsequent steps in their care. Based on the patient's medical state, background, location, and specialization, the blockchain system uses a recommendation algorithm to identify a list of potential hospitals capable of providing appropriate care. This system subsequently deploys patient contracts and hospital contracts over the blockchain, automating and enforcing agreements between the patient and the chosen hospital. The utilization of blockchain in this context offers a significant advantage in terms of secure access to patient medical records. Only authorized hospital staff can access and update patient data, ensuring confidentiality and privacy. Consequently, this facilitates seamless communication and collaboration among healthcare professionals, optimizing critical workflows. Furthermore, the proposed system's security is reinforced by filters based on probabilistic data structure algorithms, adding an extra layer of protection to all authentications.

As blockchain technology continues to evolve, its integration with advancements in IoT holds great promise for enhancing healthcare outcomes by ensuring the timely delivery of appropriate care to each patient. In conclusion, this chapter highlights the striking potential of integrating IoT, blockchain, and AI to develop a highly effective framework for managing patient medical emergencies. As the field of healthcare continues to advance, further exploration and implementation of these technologies holds tremendous promise to transform emergency care and revolutionize healthcare delivery.

REFERENCES

1. Bell, K. M. (2008). HHS national alliance for health information technology (NAHIT). *Report to the Officer of the National Coordinator for Health Information Technology on Defining Key Health Information Technology Terms*. USA, 1–40.
2. Vyas, S., Shabaz, M., Pandit, P., Parvathy, L. R., & Ofori, I. (2022). Integration of artificial intelligence and blockchain technology in healthcare and agriculture. *Journal of Food Quality*, 2022.
3. Chamola, V., Goyal, A., Sharma, P., Hassija, V., Binh, H. T. T., & Saxena, V. (2022). Artificial intelligence-assisted blockchain-based framework for smart and secure EMR management. *Neural Computing and Applications*, 1–11.
4. Tagde, P., Tagde, S., Bhattacharya, T., Tagde, P., Chopra, H., Akter, R., . . . Rahman, M. H. (2021). Blockchain and artificial intelligence technology in e-Health. *Environmental Science and Pollution Research*, 28, 52810–52831.
5. Dakhane, A., Waghmare, O., & Karanjekar, J. (2020). Blockchain AI framework for healthcare records management constrained goal model. *International Research Journal of Engineering and Technology*, 10(2).
6. Xie, Y., Lu, L., Gao, F., He, S. J., Zhao, H. J., Fang, Y., . . . Dong, Z. (2021). Integration of artificial intelligence, blockchain, and wearable technology for chronic disease management: A new paradigm in smart healthcare. *Current Medical Science*, 41, 1123–1133.
7. Fatoum, H., Hanna, S., Halamka, J. D., Sicker, D. C., Spangenberg, P., & Hashmi, S. K. (2021). Blockchain integration with digital technology and the future of health care ecosystems: Systematic review. *Journal of Medical Internet Research*, 23(11), e19846.
8. Omrčen, L., Leventić, H., Romić, K., & Galić, I. (2021, September). Integration of blockchain and AI in EHR sharing: A survey. In *2021 International Symposium ELMAR* (pp. 155–160). UK: IEEE.
9. Bloom, B. H. (1970). Space/time trade-offs in hash coding with allowable errors. *Communications of the ACM*, 13(7), 422–426.
10. Liu, Y., Du, Y., Zhang, Y., Li, Y., Cyril, L., Miao, C., . . . Tian, Z. (2022). A blockchain-based personal health record system for emergency situation. *Security and Communication Networks*, 2022.
11. Fan, B., Andersen, D. G., Kaminsky, M., & Mitzenmacher, M. D. (2014, December). Cuckoo filter: Practically better than bloom. In *Proceedings of the 10th ACM International on Conference on emerging Networking Experiments and Technologies* (pp. 75–88). Japan.
12. Kaur, D., Singh, B., & Rani, S. (2023). Cyber security in the metaverse. In *Handbook of Research on AI-Based Technologies and Applications in the Era of the Metaverse* (pp. 418–435). USA: IGI Global.
13. Kataria, A., Puri, V., Pareek, P. K., & Rani, S. (2023, July). Human activity classification using G-XGB. In *2023 International Conference on Data Science and Network Security (ICDSNS)* (pp. 1–5). Tiptur: IEEE.
14. Puri, V., Kataria, A., Rani, S., & Pareek, P. K. (2023, September). DLT based smart medical ecosystem. In *2023 International Conference on Network, Multimedia and Information Technology (NMITCON)* (pp. 1–6). Raichur: IEEE.
15. Rani, S., Kumar, S., Kataria, A., & Min, H. (2023). SmartHealth: An intelligent framework to secure IoMT service applications using machine learning. *ICT Express*, 10(2), 420–425.
16. Singh, H., & Rani, S. (2023). Flex sensor integrated smart strap to verify correct wearing of the face mask. *IEEE Sensors Journal*, 24(2), 2020–2027.
17. Rani, S., Kaur, J., & Bhambri, P. (2023). Technology and gender violence: Victimization model, consequences and measures. In *Communication Technology and Gender Violence* (pp. 1–19). Cham: Springer International Publishing.

Index

S

T

V

W